AF772469

Noah's Flood Happened 3520 B.C.

Gerald E. Aardsma, Ph.D.

ARP
Aardsma
Research & Publishing

©2015 by Gerald E. Aardsma

All rights reserved. No part of this publication may be reproduced, stored in a retrieval system or transmitted in any form or by any means: electronic, electrostatic, magnetic tape, mechanical, photocopying, recording or otherwise, without prior written permission of the publisher.

Published by:
Aardsma Research & Publishing
412 N Mulberry St
Loda, Illinois 60948-9651

www.BiblicalChronologist.org

Printed in the United States of America

Library of Congress Control Number: 2015904508

ISBN 978-0-9647665-7-0

Contents

List of Figures

List of Tables

To
Helen
my friend, companion, assistant, wife
these past four decades.
"Many daughters have done nobly,
But you excel them all."

ACKNOWLEDGMENTS

This old, fallen world tests all relationships by fire. But if a man is fortunate, advancing years may leave to him some of proven love and loyalty. Where others, glancing in, may perceive only smoking, charred remains, time has revealed to him hearts of gold. These he knows he may call on to help share his burdens, as I have done with the construction of this book.

Early drafts were proofed by my wife, Helen, my daughters Jennifer Hall and Rachel Contreras, and my daughter-in-law Esther Aardsma. Rachel and Esther went on to compile the index. Jennifer's husband Steve provided graphic design and artwork, including the cover design.

The great burden of smoothing rough edges of all sorts in later drafts was courageously shouldered and diligently prosecuted by Tom Godfrey.

The rough edges that remain are a monument to my obstinacy.

PREFACE

While I am fundamentally an independent research scientist, not a communicator, from time to time I feel an urge to communicate some of the results of my research to others. I am prodded in this direction by a sense of responsibility, especially toward fellow scientists. I am a physicist, with emphasis on physical dating methods such as radiocarbon. My special concern has been to harmonize secular and biblical chronologies of earth history. This places my research at the interface of science and the Bible. Discoveries at this interface can greatly impact a person's worldview. I have always felt that if someone else were doing this research, I would certainly want to be informed of their findings. A simple application of the Golden Rule says I should try to inform others.

But a bit of a sinking feeling always dampens my enthusiasm when I undertake one of these communication projects. Experience has shown repeatedly that most individuals, scientist or otherwise, are not very receptive to new results challenging their accustomed way of thinking.

This observation extends far beyond my own experience. In the September 1993 issue of *The Quarterly Review of Biology*, Professor Müller-Hill of the Institut für Genetik der Universität zu Köln in Germany observed:[1]

> In science and elsewhere there are two types of truth: (1) The truth everybody already knows, and (2) the truth that is not yet discovered. Most persons deal in science, as elsewhere, with the first type of truth. Most scientists just analyze another homologous system, and thus simply produce more of

[1]Benno Müller-Hill, "Science, Truth, and Other Values," *The Quarterly Review of Biology* 68.3 (September 1993): 399.

> the same. The second type of truth is different. At first it looks too bizarre to be true, and it may be as dangerous as fire. If you are not clever it may destroy you.

The pages of history are replete with instructive instances of this phenomenon. Galileo is probably the most celebrated example.

Galileo Galilei was a professor of mathematics in his mid-forties when, toward the latter part of the first decade of the seventeenth century after Christ, he learned about a newly invented instrument which was said to make distant objects look much closer. It was a spyglass, a first primitive telescope, the earliest forms of which were not too effective, with a magnification of only three or four. Galileo quickly built his own spyglass and proceeded to make improvements on its design until he had produced a twenty-powered spyglass. He soon used this instrument to view the moon—and he was immediately thrown into a great conflict with the wisdom of his day, and, indeed, with wisdom of great antiquity, by what he saw.

A geocentric cosmology prevailed at the time, as it had for a very long time before. According to this view of the physical universe, the heavens were the realm of God, and the earth was the realm of men. Since the heavens were the realm of God, they were regarded as necessarily perfect and unchanging. And this conception included the idea that the heavenly bodies, such as the moon, were all geometrically perfect spheres.[2]

> According to the then prevailing geocentric cosmology of Aristotle, the heavens were perfect and unchanging, and heavenly bodies were perfectly smooth and spherical. The large spots visible on the Moon to the naked eye were usually explained away by ad hoc devices. One could, for instance, postulate that parts of the perfectly smooth Moon absorbed and then emitted light differently from other parts.

Now I hope you do not side with the "chronological snobs" (C. S. Lewis' term, as I recall, for those who look down their noses at others who have lived before them, supposing the advancement in knowledge which they are privileged to partake of through no merit of their own is evidence of their intrinsic superiority) and regard everybody who lived back in

[2]Galileo Galilei, *Sidereus Nuncius*, translated with introduction, conclusion, and notes by Albert Van Helden (Chicago: The University of Chicago Press, 1989), 10–11.

Galileo's day as foolish for believing such things. Galileo's contemporaries were not lacking in intelligence—they were really no different from people today in that respect. Next time you are out of doors on a moonlit night, take a long look at the moon with your unaided eyes, and see if you can discern any deviation from perfect smoothness in its orb. And then see how successfully you can answer the question of why God should have created the moon with the pocked and pitted surface we have come to understand it actually possesses.

The view held by Galileo's contemporaries was of very ancient and respectable lineage. It was theologically satisfying. And it was empirically attested by every person's own eyes—until Galileo sighted his spyglass on the moon and became the first man ever to behold its majestic mountains and sunken craters.

In a letter dated January 7, 1610, Galileo wrote:[3]

> ... it is seen that the Moon is most evidently not at all of an even, smooth, and regular surface, as a great many people believe of it and of the other heavenly bodies, but on the contrary it is rough and unequal. In short it is shown to be such that sane reasoning cannot conclude otherwise than that it is full of prominences and cavities similar, but much larger, to the mountains and valleys spread out over the Earth's surface.

Later in 1610 Galileo published his discovery in a little book called *Sidereus Nuncius*, together with the further startling discovery that Jupiter was orbited by four moons of its own—an observation which conflicted severely with the geocentric cosmology of his day, which held that the earth was the single center of rotation in the universe. Besides this publication, he worked feverishly to produce other telescopes of high quality so other scientists could check his observations. And he wrote letters and gave lectures and carried out personal visits to eminent scientists of his day replete with late-night demonstrations of his observations.

It is well known how Galileo's discoveries were ultimately received by the religious establishment of his day—how he spent the latter years of his life under house arrest. Not so well publicized is how his discoveries were treated by other *scientists* of his day.

---

[3]Galileo Galilei, *Sidereus Nuncius*, translated with introduction, conclusion, and notes by Albert Van Helden (Chicago: The University of Chicago Press, 1989), 11.

In April 1610 Galileo visited an astronomer of international reputation, Giovanni Antonio Magini, bringing his spyglass with him. He evidently demonstrated the instrument for a gathering of local scholars and allowed it to be thoroughly investigated by them. Their appraisal was chronicled a few days later by Martin Horky, a young associate of Magini, in a letter to the now famous astronomer Johannes Kepler (eight years younger than Galileo):[4]

> Galileo Galilei, the mathematician of Padua, came to us in Bologna and he brought with him that spyglass through which he sees four fictitious planets [*i.e., moons of Jupiter*]. On the twenty-fourth and twenty-fifth of April I never slept, day and night, but tested that instrument of Galileo's in innumerable ways, in these lower [earthly] as well as the higher [realms]. On Earth it works miracles; in the heavens it deceives, for other fixed stars appear double. Thus, the following evening I observed with Galileo's spyglass the little star that is seen above the middle one of the three in the tail of the Great Bear, and I saw four very small stars nearby, just as Galileo observed about Jupiter. I have as witnesses most excellent men and most noble doctors, Antonio Roffeni, the most learned mathematician of the University of Bologna, and many others, who with me in a house observed the heavens on the same night of 25 April, with Galileo himself present. But all acknowledged that the instrument deceived. And Galileo became silent, and on the twenty-sixth, a Monday, dejected, he took his leave from Mr. Magini very early in the morning. And he gave no thanks for the favors and the many thoughts, because, full of himself, he hawked a fable. Mr. Magini provided Galileo with distinguished company, both splendid and delightful. Thus the wretched Galileo left Bologna with his spyglass on the twenty-sixth.

To his great credit, Kepler disregarded Horky's appraisal and, true to his own nature, accepted Galileo's observations enthusiastically. But

[4]Galileo Galilei, *Sidereus Nuncius*, translated with introduction, conclusion, and notes by Albert Van Helden (Chicago: The University of Chicago Press, 1989), 92–93. The italicized words in square brackets are mine, non-italicized words in square brackets are in the original.

most others were less generous.[5]

> [Galileo] also received many letters in which objections to his discoveries were put forward, and answering them all was a frustrating business:
>
> > It is true that their reasons for mistrust are very frivolous and childish, since they persuade themselves that I am so rash that in testing my instrument a hundred thousand times on a hundred thousand stars and other objects, I have not known, or been able to recognize, those deceptions that they think they have recognized without ever having seen the instrument; or else, that I am so stupid that without any need I have wished to compromise my reputation and to ridicule my Prince.

While Galileo's discoveries were not well received by many of the leading men of his day, we must not judge these individuals harshly in this, for it is too true, as Professor Müller-Hill has pointed out above, that new truth often "looks too bizarre to be true."

I do not pretend to possess the genius of a Galileo, but I have, like Galileo, had the joy of discovering something which has previously been hidden from human understanding. I have, like him, exerted myself from time to time to communicate what I have discovered. And I have, like him, had a limited reception.

Several years ago, for example, I submitted a paper to a well-known, mainstream secular scientific journal arguing, as I had previously done in my own bi-monthly newsletter, *The Biblical Chronologist*,[6] that the chronology of laminated sediments from Elk Lake, Minnesota had been misinterpreted.[7] The original researchers had interpreted a thick section of anomalous sediment as spanning a six-hundred-year interval. I pointed out that several sets of data from Elk Lake contradicted this interpretation. I argued that to avoid the contradictions posed by these datasets it was necessary to interpret this anomalous section as due to a single brief

---

[5] Galileo Galilei, *Sidereus Nuncius*, translated with introduction, conclusion, and notes by Albert Van Helden (Chicago: The University of Chicago Press, 1989), 100.

[6] http://www.BiblicalChronologist.org/products/archives/index.php

[7] Gerald E. Aardsma, "Noah's Flood at Elk Lake," *The Biblical Chronologist* 2.6 (November/December 1996): 1–13.

episode of intense deposition. I suggested that careful radiocarbon measurements should be able to resolve the true chronology of the anomalous section and that such a check should be carried out to settle the issue.

I would have liked to have added that the anomalous section dated to the time of Noah's Flood, implying that Noah's Flood was, in fact, the explanation of the anomalous section of sediment in question. But I knew that mention of Noah's Flood in a way which treated it as real history would almost certainly guarantee rejection of the paper by the journal editor prior even to the usual peer review process.

Now Noah's Flood is, in fact, real history, as this volume will amply demonstrate. But the world, and most especially the academic segment of the world, has not been very receptive to this fact to the present time. They can hardly be blamed for this. For several centuries now, the tide of scientific discovery has seemed to be all contrary to the biblical book of Genesis, from which the Western World has traditionally gained its knowledge of Noah's Flood. And the Bible/science interface addressing the Flood and other issues from the early chapters of Genesis has tended to be overrun from both sides with zealots, crackpots, and impostors.

Prudence demanded that I attempt this communication in two stages: (1) establish the fact that there was a big chunk of sediment at Elk Lake which was deposited in a single brief episode, rather than over the course of 600 years, and, once that was published, (2) point out in a follow-up paper that this single brief episode was synchronous with the date of Noah's Flood calculated by modern biblical chronology, and suggest that Noah's Flood provided the only reasonable explanation of the anomalous sediments.

I heard back from the editor five months later. The paper had been rejected following peer review. But the editor cordially offered, "If you can make a strong case, based on further evidence or reasoning, that would refute the referees' assessments, we would welcome hearing from you."

The referees—including the scientist who had interpreted the anomalous section of sediments in terms of 600 years of slow deposition to begin with—did not deny that there were problems with the 600-year interpretation. But to accept the idea that the anomalous section of sediments was deposited during a single brief episode, and allow the paper to be published, both referees felt, as the editor summarized, that "a reasoned explanation of the rapid accumulation of the layered sediments is required."

Well, they had me there. The chronology issues should have been able to be treated by themselves on the basis of chronological data alone. But the referees were demanding to see a physical model for rapid accumulation of layered sediments at Elk Lake before they would allow the chronological issues even to be surfaced.

What they were requiring meant that I could not split the communication in two. The only way I could begin to give a "reasoned explanation of the rapid accumulation of the layered sediments" at Elk Lake was to bring Noah's Flood into the picture. And I seriously doubted their ability to give that idea a fair hearing.

I pondered my dilemma off and on for a year, but I could find no solution. Eventually I decided I had no choice. I had to state in the paper that the anomalous section of sediment found reasonable explanation only within the context of a historical Flood. The alternative was to forget trying to communicate the implications of Elk Lake sediments to my scientific colleagues altogether—which was fine with me; I had several plates full of other interesting discoveries in progress which were more than enough to keep me occupied. But there was still the Golden Rule to reckon with. The door hadn't been entirely slammed in my face yet. Who could say but that the editor and referees might rise above history's norm in this one instance...? (I tend toward an overriding optimism about people. Life has been trying to beat this out of me, but still it persists.)

So I enlarged the original paper to include an explanation of the anomalous sediments in terms of Noah's Flood. I showed that the secular date for this anomalous sediment from Elk Lake coincided with the biblical date of Noah's Flood. And I summarized the archaeological and geophysical evidence for the historical reality of Noah's Flood which I had found up to that time. The thrust of the original paper was retained (i.e., there is a chronological problem with the Elk Lake sediments which needs to be resolved). And most of the original discussion was retained. In short, it was the same paper with an added explanation of how the anomalous laminated sediments might have been rapidly deposited, to meet the demands of the referees.

Well, at least I didn't need to wait five months for a reply this time. A very definitive rejection was forthcoming in just six weeks. And this time the editor did not volunteer that I should write again with any further thoughts.

Since the editor had sent the original paper out for review, he had had little choice but to send the enlarged version out also—despite its reference to a historically real Noah's Flood. That is why it took all of six weeks for the reply. He chose a different set of two referees for the enlarged paper. Their comments were not models of scientific objectivity. I will spare you the full treatment; here is a sample:

> This paper is not science, not even pseudo-science. Its approach to varve analysis requires enormous faith and not any factual grasp of reality. I wouldn't even call this paper speculation. There is no evidence presented that can be used speculation.[8] There appears to be such a rampant desire to show the existence of a catastrophic flood that the paper is blind to what can and cannot be substantiated or what is even realistic. There is NO support for any of the author's assumptions, which are necessary for any part of this house of cards to have credibility.

Well, I suppose I should have known better. Not only is human nature what it is, but the truth about Noah's Flood was way too big and looked much too bizarre to be communicated piecemeal, even back then.

So I am trying again, this time in a book of over 300 pages.

This book has been under construction for over five years. Some of the material in it has previously appeared within the pages of *The Biblical Chronologist*, which takes it back as much as another 15 years prior to that. *The Biblical Chronologist*, or *BC* for short, records much of my own personal quest to understand the truth about the Flood as that quest was unfolding. For this reason, old *BC* articles on this topic are necessarily fragmentary, somewhat disjointed, and sometimes off on a wrong trail altogether. The present volume seeks to put things in proper order, to correct as necessary, and to add new material as needed to catch the quest up to date.

Though I am a chronologist, this book is not about chronology. It is about Noah's Flood. It is meant to show that the Flood is real history, and to show as much as possible what Noah's Flood was really like. And it

[8] That is how this sentence appears in the original. I do not know how to emend it since I am not sure what it was meant to say.

is meant to awaken earth's occupants, especially its scientific community, to the fact that the hazard of the Flood is not merely a one-time thing of the past.

In the following pages, the phenomenon of scientists struggling to force their data, which, unbeknownst to them, are due to Noah's Flood, into an explanation devoid of Noah's Flood will be repeatedly encountered. It is not a pretty sight. I hope you will not be too hard on these scientists. It is not easy to think outside the accepted paradigm in any community, even when your data refuse to cooperate with the accepted paradigm. And it is at least doubly hard to do so when to face up to the simple truth of your data is likely to cost you your reputation and your livelihood.

Gerald E. Aardsma
December 12, 2013
Loda, IL

Part 1

The Historicity of Noah's Flood

CHAPTER ONE

NOAH'S FLOOD TODAY

William G. Dever:[1] Most biblical scholars regard most of the stories in Genesis as myths.

...

Hershel Shanks:[2] It's true, I think, that the first 11 chapters of Genesis would be regarded as myths—the creation stories, the story of Noah and the flood ...[3]

The mainstream view in academia today is that the biblical narrative of the Flood is a myth. This view is mistaken.

Opposite to this view is that held by a significant segment of evangelical Christianity today. Through the efforts of such "creation-scientists" as the late Dr. Henry Morris, the idea has been popularized that the Flood was an earth-shattering tectonic cataclysm responsible for most of earth's sedimentary rocks and their entombed fossils.[4] This view is also mistaken.

Large advances have been made in several fields of study pertinent to the Flood in the past several decades. When taken together, these advances reveal both that the Flood was not a myth, and that, while it was

[1]Professor of Near Eastern Archaeology and Anthropology, University of Arizona; previously Director of W. F. Albright Institute of Archaeological Research, Jerusalem.

[2]Editor of *Biblical Archaeology Review.*

[3]Hershel Shanks, "Is This Man a Biblical Archaeologist?" *Biblical Archaeology Review* 22.4 (July/August 1996): 62.

[4]John C. Whitcomb, Jr. and Henry M. Morris, *The Genesis Flood* (Philadelphia: The Presbyterian and Reformed Publishing Company, 1961), 242–243.

of global proportions, it was not an earth-shattering cataclysm. These advances reveal remains of civilizations which perished in the Flood—pre-Flood peoples' houses, their tools, their burial practices, their artistic abilities, and much more. They also clarify the true nature of the Flood—its geographical extent, where the water for the Flood came from, where all the water went to when the Flood was over, how deep the water was at various times throughout the Flood, what the underlying physical mechanism of the Flood must have been, and, again, much more.

One field which has made enormous strides of great importance to our understanding of the Flood is biblical archaeology. Large increases in the number of archaeological excavations in recent decades have brought forth a wealth of archaeological data which previous generations knew nothing about. Biblical archaeology has now opened a wide door into the same ancient past we read about in Genesis.

Another field which has made great progress is radiocarbon dating. This dating method was calibrated via dendrochronology (i.e., the tree-ring counting dating method) in the 1980s. The calibration covers the entire timespan of interest to the biblical historical narrative. It removes doubts and uncertainties about the constancy of the production of radiocarbon atoms in the upper atmosphere and other such factors, rendering earlier criticisms of the radiocarbon dating method obsolete. Modern tree-ring-calibrated radiocarbon dating allows a truly objective, universal chronology of the ancient past to be constructed for the first time ever.

Other fields have made remarkable advances in the past several decades, contributing in their own way to our ability to understand the ancient past more correctly as it touches on the Flood. This will become increasingly apparent in the following pages, but there is one other field which must be mentioned here. This is the very old discipline of biblical chronology. It was a most unexpected yet vital advance in this field (Chapter 3) in the early 1990s which touched off the modern discovery of the true, historical Noah's Flood, overturning previous views.

CHAPTER TWO

HOW NOT TO FIND NOAH'S FLOOD

Success in finding the true, historical Noah's Flood has everything to do with *chronology*.

Chronology is, as has often been said, the backbone of history. This means that to get history right, chronology must first be gotten right. And if it is biblical history which is being dealt with, then it is biblical chronology which must first be gotten right.

The following two examples, one old and one new, illustrate this point.

The Woolley "Flood"

Archaeologist Leonard Woolley excavated at Ur from 1922 to 1934. He made many important archaeological discoveries in the process. He also claimed to have found Noah's Flood at Ur. He described the discovery of his "Flood" as follows:[1]

> In the lowest levels of potsherds the Erech ware was mixed with and finally replaced altogether by al 'Ubaid pottery, hand-made without the wheel, the product of the early and still barbarous villagers who first settled at Ur. Then we came to the flood deposit, eleven feet of clean silt, disturbed only

[1]Sir Leonard Woolley, *Spadework in Archaeology* (New York: Philosophical Library, 1953), 105-106.

> by a few graves dug into it by the late al 'Ubaid people whose pottery we had found; silt left piled up against the mound whereon the primitive town had stood by an inundation that must have overwhelmed all the low-lying villages of the river valley and destroyed what for those people was the world. Under the eleven-foot stratum lay the ruins of the houses in which had lived the antediluvian inhabitants of the Lower Town; they had, we may suppose, taken refuge on the mound, the Inner City, and from its walls watched their homes disappear beneath the muddy waters of the flood. This was the flood, coming in the latter part of the al 'Ubaid period, which the ancient Sumerians regarded as the outstanding disaster in their country's history, and out of the historic fact grew the legend which in course of time the Hebrew people incorporated in their own sacred writings and handed on to us today, the story of Noah's Flood.

There is little which can be commended in Woolley's method of identifying his "Flood." His evidence consists of "eleven feet of clean silt" among stratified archaeological deposits at Ur. This is not much to go on, to be sure.

Woolley failed to show any extension of this silt layer to any other geographical location, including even other nearby cities. And not only did his "Flood" fail to cover high mountains as the biblical narrative stipulates,[2] it failed to cover even the mound on which the Inner City stood.

Not that any of this made much difference to Woolley. Having prejudged the Bible's narrative of the Flood to be mere legend, what Genesis had to say about the Flood was obviously neither here nor there with him. This left the field wide open for the "Flood" to be whatever Woolley chose to imagine it to be.

But it is mainly Woolley's treatment of chronology which is of interest at present, not his general methodology or philosophy. What biblical chronological data does he mention? None. What secular chronological data does he mention? None. No dates show up anywhere in Woolley's discussion of his "Flood" discovery.

---

[2]Genesis 7:19.

This is really very surprising. Surely there have been many floods within the Tigris and Euphrates river basins throughout the thousands of years of the ancient past. On what basis should the claim be accepted that this particular flood deposit at Ur was due to Noah's Flood? Surely an absolutely minimal requirement, even within Woolley's cavalier methodology, is some sort of assurance that this silt was deposited at the same time the event underlying the "legend" of Noah's Flood should be expected to have taken place. But there is no such assurance.

It is, I think, very common to assume that since an individual is trained in archaeology, that individual is automatically competent also in chronology. This assumption is, in fact, false. Archaeology and chronology are two separate fields of study. Training in one of these fields does not confer expertise in both.

In the present case, Woolley's silence in regard to the chronology issues involved with his claimed discovery of the Flood most certainly did not flow from an automatic expertise with chronology. Rather it flowed from a profound chronological ignorance of the remote period he was excavating, endemic to the archaeology of his day.

Not surprisingly, Woolley's "Flood" has generally been repudiated by liberal and conservative alike. For example, the accomplished archaeologist, G. Ernest Wright, wrote some time ago:[3]

> This archaeologist [Woolley] discovered a water-laid deposit at Ur, some 10 feet (3 meters) thick, dating from the middle of the Ubaid period of the 4th millennium B.C. This, he claimed, was conclusive evidence of the Flood. Yet in only two of the five pits that Woolley dug to virgin soil did he find the "flood" layer. This evidence suggests that the flood in question did not cover the whole city; and we know that it made no break in the continuity of culture there.
>
> Other cities in the Mesopotamian river valley, notably Kish, Fara, and Nineveh, also show flood layers, though none of them can be closely correlated in time. On the other hand, the excavators have found no such layer in Gilgamesh's own city of Erech (Warka). In other words, the Mesopotamian "flood" evidence is that of purely local inundations of the

[3]G. Ernest Wright, "The Flood," *The Encyclopedia Americana – International Edition* (Danbury, Connecticut: Grolier Incorporated, 1993), 415.

Tigris and the Euphrates.

Chronology is the backbone of history. Archaeological reconstructions of the past which fail to give careful consideration to chronology are not just error-prone, they are recipes for guaranteed failure.

The Ryan and Pitman "Flood"

Marine geologists Bill Ryan and Walter Pitman of Columbia University coauthored *Noah's Flood: The New Scientific Discoveries about the Event that Changed History*, published in 1998. They proposed that a catastrophic flooding of the Black Sea was the historical reality behind Noah's Flood.

Why they should pin their Black Sea flood on Noah is harder to understand even than in the case of Woolley. Connections between the biblical Flood and Ryan and Pitman's "Flood" are almost nonexistent.

- Noah's Flood, according to modern biblical chronology, happened 3520 B.C. (See the next chapter.) Ryan and Pitman's "Flood" happened 5600 B.C.—some say 7400 B.C.
- Noah's Flood covered high mountains. Ryan and Pitman's "Flood" raised the surface of the Black Sea by 80 meters—some say 5 meters.
- Noah's Flood took the building of an ark to escape. Ryan and Pitman's "Flood" was survived by walking away.
- Noah's Flood drowned just about everybody. Ryan and Pitman's "Flood" drowned just about nobody.
- Noah's Flood landed the ark in the mountains of Ararat. Ryan and Pitman's "Flood" raised nothing any higher than the surface of the Black Sea.
- Noah's Flood dried up, and the waters returned to normal. Ryan and Pitman's "Flood" still holds its flooded height today.

On the positive side, there is the fact that Noah's Flood and Ryan and Pitman's "Flood" did both involve water.

I will not bother to delve into this "Flood." It is discussed in great detail on the Internet for any who may be interested. But notice that it does share some similarities with Woolley's "Flood" of over half a century ago—its nearly non-existent evidential basis, its headline-grabbing announcement, its demise in the face of further scholarly inquiry, and its suicidal neglect of chronology.

Conclusion

The main take-home lesson with these examples is a simple one. It is possible to go on inventing "Floods" indefinitely. All it takes is a little data from just about any period of time and an unrestrained imagination.

To stay out of this endless world of make-believe and home in on the singular truth—the real historical Noah's Flood—there must be an insistence, first and foremost, on rigorous adherence to the principles of sound chronology, both biblical and secular. Chronology specifies *when* in history an event took place, thus eliminating from consideration all other events at all other dates. Any flood which does not date to the time of Noah is not Noah's Flood.

CHAPTER THREE

DATING NOAH'S FLOOD

The Bible, which tells about Noah and about the Flood which extinguished civilization in his day, also provides the raw chronological data needed to learn when in the past the Flood should be looked for. The proper handling of biblical chronological data is the business of the discipline of biblical chronology. This discipline must be consulted to convert the Bible's raw chronological data into a B.C. date for the Flood.

Traditionally, Noah's Flood has been dated by biblical chronologists (men like Bishop Ussher) to the middle of the third millennium B.C. This date has been untenable for quite some time.

The account of the Flood found in Genesis 6–9 specifies flood waters deep enough to submerge high mountains[1] for 150 days[2] with consequent near-extermination of pre-Flood civilization.[3] The archaeological record of the ancient Near East reveals nothing suitable to this in the middle of the third millennium B.C. In actual fact, civilizations throughout the Near East show basic continuity all through the third millennium B.C. This is one of the reasons underlying modern academia's view that the Flood is a myth.

But the Flood is not a myth. The Bible is not mistaken about the historical reality of the Flood. The problem is that old-time biblical chronologists were mistaken about the date of the Flood.

If you look for the Flood in the middle of the third millennium B.C., where old-time biblical chronologists have traditionally placed it, you will

[1]Genesis 7:19.
[2]Genesis 7:24.
[3]Genesis 7:21–23.

not find it, no matter how hard or long you search for it there. This is because the Flood did not happen in the middle of the third millennium B.C. It happened in the middle of the *fourth* millennium B.C.

A Lost Millennium in 1 Kings 6:1

What old-time biblical chronologists did not catch was that a primary biblical number, needed to compute the date of the Flood, had been corrupted by a rare copy error in the distant past. Four Hebrew letters had been accidentally dropped from the ancient text of 1 Kings 6:1, causing an original 1480 years to be truncated to just 480 years in all extant manuscripts (Figure 3.1).[4]

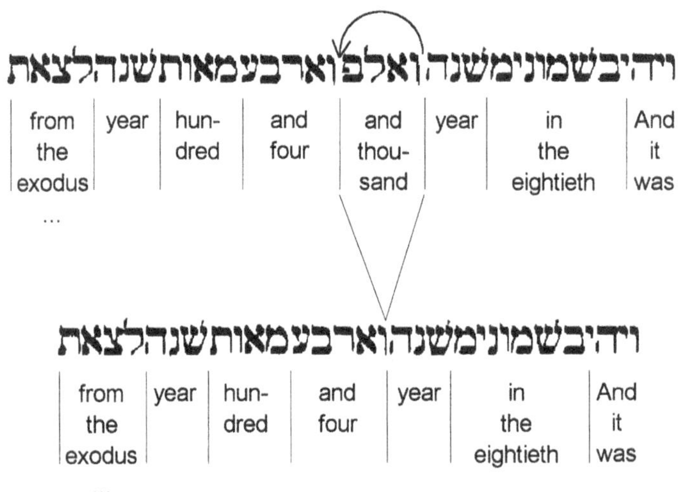

Figure 3.1: *A rare copy error in the remote past has resulted in the loss of one thousand years from the present text of 1 Kings 6:1. The proposed original text is shown on the upper line (read from right to left); the present text is shown on the lower line. The arrow indicates how, due to a repeated letter sequence, the scribe's eye might accidentally have skipped over "and thousand" in the process of copying the original.*

[4]Gerald E. Aardsma, *A New Approach to the Chronology of Biblical History from Abraham to Samuel*, 2nd ed. (Loda IL: Aardsma Research and Publishing, 1993). www.BiblicalChronologist.org.

When the accidentally dropped "and thousand" is restored to 1 Kings 6:1, dates of biblical events prior to the birth of Eli (about 1200 B.C.)—including the Flood—are moved back one thousand years relative to traditional thinking. The validity of this restored reading is evidenced by the harmony between the biblical narrative and secular studies which it consistently produces.[5]

A surprising oversight—yet easily understood

In one sense, it is surprising that old-time biblical chronologists did not catch this copy error, for they were capable scholars, and there are some clear hints within the pages of the Bible itself that something is wrong with the present "480 years" of 1 Kings 6:1. For example, 1 Kings 6:1 says this "480 years" covers the period of time from the Exodus to the fourth year of Solomon's reign. Yet other biblical chronological data (Table 3.1, page 38) yield a *minimum* duration of 600 years for this same time interval.[6]

But in another sense, the failure of old-time biblical chronologists to catch the copy error in 1 Kings 6:1 is easily understood. Though the biblical hints are there, they are easily muted. It is easy, for example, to explain away the problem presented by the Table 3.1 data by the postulate that the tenure of the various judges chronicled in the book of Judges must have overlapped—despite the fact that the Judges narrative gives the average reader every impression of being deliberately chronological and consecutive. In consequence, it took the amassing of great quantities of biblical archaeology data in modern times, at well-known Holy Land sites such as Jericho and Ai, to make the problem with 1 Kings 6:1 unavoidably obvious.

The Proper Date of the Flood

Table 3.2 (page 39) shows the biblical data used to obtain a B.C. date for the Flood once the missing thousand years have been restored to 1 Kings 6:1.

---

[5]See, for example, Gerald E. Aardsma, *The Exodus Happened 2450 B.C.* (Loda IL: Aardsma Research and Publishing, 2008). www.BiblicalChronologist.org.

[6]Gerald E. Aardsma, *A New Approach to the Chronology of Biblical History from Abraham to Samuel*, 2nd ed. (Loda IL: Aardsma Research and Publishing, 1993), 27. www.BiblicalChronologist.org.

Item	Reference	Number	Total
wilderness wanderings	Numbers 14:33–34	40	40
Conquest to first oppression	Joshua 1:1 – Judges 3:7	>10	>50
first oppression	Judges 3:8	8	>58
Othniel's judgeship	Judges 3:11	40	>98
oppression by Eglon	Judges 3:14	18	>116
Ehud's judgeship	Judges 3:30	80	>196
oppression by Jabin	Judges 4:3	20	>216
Deborah's judgeship	Judges 5:31	40	>256
oppression by Midian	Judges 6:1	7	>263
Gideon's judgeship	Judges 8:28	40	>303
Abimelech's reign	Judges 9:22	3	>306
Tola's judgeship	Judges 10:2	23	>329
Jair's judgeship	Judges 10:3	22	>351
oppression by Ammon	Judges 10:8	18	>369
Jephthah's judgeship	Judges 12:7	6	>375
Ibzan's judgeship	Judges 12:9	7	>382
Elon's judgeship	Judges 12:11	10	>392
Abdon's judgeship	Judges 12:14	8	>400
oppression by Philistines	Judges 13:1	40	>440
Samson's judgeship	Judges 15:20	20	>460
Samson to Eli	Judges 17 and following	>10	>470
Eli's judgeship	1 Samuel 4:18	40	>510
Samuel's judgeship	1 Samuel 8:1,5	>15	>525
Saul's reign	1 Sam. 13:1 (NASB 1975)	32	>557
David's reign	2 Samuel 5:4	40	>597
Solomon's fourth year	1 Kings 6:1	3	>600

Table 3.1: Biblical chronological data yielding more than 600 years from the Exodus to Solomon's fourth year. (In three instances the biblical historical narrative does not supply the needed chronological data. These are indicated by > (greater than) signs in column 3 of the table, followed by deliberately conservative guesses at the elapsed time in each instance.)

Item	Reference	Number	Date (B.C.)
accession of Rehoboam			931±10
accession of Solomon	1 Kings 11:42	40±5	971±11
Solomon's fourth year	1 Kings 6:1	-4±0.5	967±11
Exodus from Egypt	1 Kings 6:1 (restored)	1480±5	2447±12
Israel enters Egypt	Exodus 12:40–41	430±0	2877±12
Jacob before Pharaoh	Genesis 47:1–11		2877±12
birth of Jacob	Genesis 47:9	130±5	3007±13
birth of Isaac	Genesis 25:26	60±5	3067±14
birth of Abraham	Genesis 21:5	100±5	3167±15
Abraham enters Canaan	Genesis 12:4	-75±5	3092±16
death of Terah	Acts 7:4		3092±16
birth of Terah	Genesis 11:32	205±5	3297±17
birth of Nahor	Genesis 11:24	29±5	3326±17
birth of Serug	Genesis 11:22	30±5	3356±18
birth of Reu	Genesis 11:20	32±5	3388±19
birth of Peleg	Genesis 11:18	30±5	3418±19
birth of Eber	Genesis 11:16	34±5	3452±20
birth of Shelah	Genesis 11:14	30±5	3482±21
birth of Arpachshad	Genesis 11:12	35±5	3517±21
end of the Flood	Genesis 11:10	2±0.5	3519±21

Table 3.2: Data used to compute the date of Noah's Flood.

These data yield a biblical chronology date for the end of the Flood of 3519±21 B.C.[7] Since the Flood appears, from the biblical narrative, to have lasted exactly one year (Chapter 16), the corresponding biblical chronology date of the start of the Flood is 3520±21 B.C.

Conclusion

3520 B.C. is the best current biblical chronology date for the Flood. This places the Flood in world history and prehistory in the middle of the *fourth* millennium B.C., not the middle of the *third* millennium B.C. where it has traditionally been sought (Figure 3.2).

Proof of this placement follows from the fact that it works, as subsequent chapters will show. That is, an event satisfying the biblical narrative of the Flood and its aftermath, while never found anywhere in the third millennium B.C., is immediately found to be present within the data of various secular disciplines, such as archaeology and geophysics, in the middle of the fourth millennium B.C.

[7]For a full discussion of this table see Gerald E. Aardsma, "Chronology of the Bible: 5000–3000 B.C.," *The Biblical Chronologist* 2.4 (July/August 1996): 1–5. www.BiblicalChronologist.org.

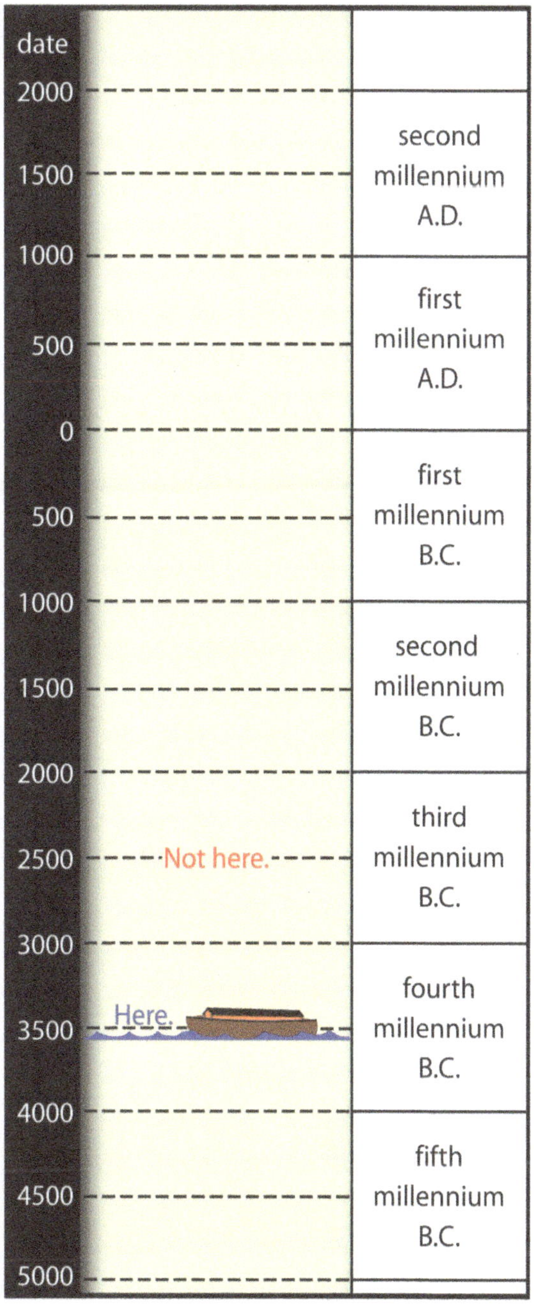

Figure 3.2: The proper chronological placement of the Flood.

CHAPTER FOUR

FINDING NOAH'S FLOOD

> And the water prevailed more and more upon the earth, so that all the high mountains everywhere under the heavens were covered. The water prevailed fifteen cubits higher, and the mountains were covered. And all flesh that moved on the earth perished, birds and cattle and beasts and every swarming thing that swarms upon the earth, and all mankind; of all that was on the dry land, all in whose nostrils was the breath of the spirit of life, died. Thus He blotted out every living thing that was upon the face of the land, from man to animals to creeping things and to birds of the sky, and they were blotted out from the earth; and only Noah was left, together with those that were with him in the ark.
>
> Genesis 7:19–23 (NASB 1975)

The biblical narrative makes extraordinary claims of the Flood and its aftermath. These claims are of a sufficiently singular nature to guarantee unambiguous identification of the Flood when they are satisfied by the real-life data of secular disciplines.

Claim #1: Size of the Flood

Noah's observations of the Flood, recorded in Genesis, including the passage quoted above, make it clear that the Flood was not a small local, or even a modest regional affair. The fact is that it is physically impossible to cover "high mountains" for months just locally or regionally. Adjacent regions will also need to be flooded if the waters are to maintain their depth. As it turns out, there *is* a way of covering high mountains for months in just one hemisphere of the earth, without flooding everything in the other hemisphere, as will be shown in subsequent chapters, so it is not necessary to conclude from this argument that what Noah experienced was a planet-wide inundation. But for the present, it is only the fact that Noah's Flood was of an extraordinary size which is of interest. The biblical text makes it clear that the true historical Noah's Flood was characterized by a geographical coverage extending far beyond local or regional boundaries.

Claim #2: Impact on Ancient Civilization

In addition to the extraordinary size of the Flood, Genesis makes the singular claim that ancient civilization was all but exterminated by the Flood.[1] The more or less complete destruction of mankind is, Genesis informs us, what the Flood was fundamentally about.[2] This claim is of first importance to the quest for the Flood in secular data. Evidence of an abrupt near-termination of civilization is the most fundamental requirement of the biblical account of the Flood.

Claim #3: Origin of Human Government

In addition, Genesis places the origin of human government—of the sort bearing responsibility to punish murderers and having authority to exercise capital punishment (more than just a council of elders, for example)—in a divine decree given to Noah following the Flood:[3]

> And surely I will require your lifeblood; from every beast I will require it. And from every man, from every man's brother I will require the life of man. Whoever sheds man's blood, by man his blood shall be shed, for in the image of God He made man.

---

[1]Genesis 7:28.
[2]Genesis 6:7, 13, 17.
[3]Genesis 9:5,6; NASB 1975.

The reason God sent the Flood, Genesis tells us, was that civilization prior to the Flood was characterized by violence.[4] There was in existence at that time no human institution with the job of keeping the peace. No institution of human government responsible for maintaining law and order within society had yet been invented. In consequence, it appears that anarchy reigned. Wanton bloodshed seems to have been commonplace, typified in Genesis by the bragging speech of Lamech to his two wives: "I have killed a man for wounding me; and a boy for striking me."[5]

The problem leading to the Flood, Genesis informs us, was anarchy and violence. The cure, given by God to mankind following the Flood, was human government having responsibility and authority to exercise capital punishment. Thus the Bible's recitation of remotest history leads to an expectation that this sort of human government will be a part of human societies only following the Flood.

Conclusion

A plain-sense reading of the biblical account of the Flood reveals three telltale signatures of the Flood. The first is the Flood's global-scale proportions. The second is its destruction of human populations. And the third is the inception of capital punishment-wielding human government immediately following the Flood. While these three signatures do not exhaust the possibilities presented by the biblical narrative for identifying the Flood in secular disciplines, they seem certainly sufficient for that purpose.

On a local scale, loss of human populations, so that a hiatus in occupation follows, is something which probably every archaeologist is familiar with. Wars, disease, shifting economies, and climate change are all reasons why a once inhabited region may become uninhabited. But while local loss of human populations may be relatively common, loss of human populations on the vast geographical scale of interest to the biblical Flood is exceedingly rare. This, by itself, guarantees that not many events suitable to the Flood will be found in our planet's past.

Similarly, the inception on Earth of the idea of an institution of human government having authority and responsibility to exercise capital

[4]Genesis 6:13.

[5]Genesis 4:23; NASB 1975.

punishment against murderers must be regarded as at least a very rare, if not indeed singular event.

There is good reason, therefore, to be confident that the true, biblical Noah's Flood has been found within the data of secular disciplines when: (1) on a global or hemispheric scale (2) loss of human populations is observed (3) with the onset of government wielding capital punishment following on the heels of that depopulation event.

And that is precisely what *is* found in the middle of the fourth millennium B.C.—where the Bible's own chronology says it should be found. And this combination is found *only* there.

CHAPTER FIVE

NOAH'S FLOOD IN SOUTH MESOPOTAMIA

South Mesopotamia (southeastern Iraq; Figure 5.1) is the geographical setting of the biblical historical events leading up to the Flood. The

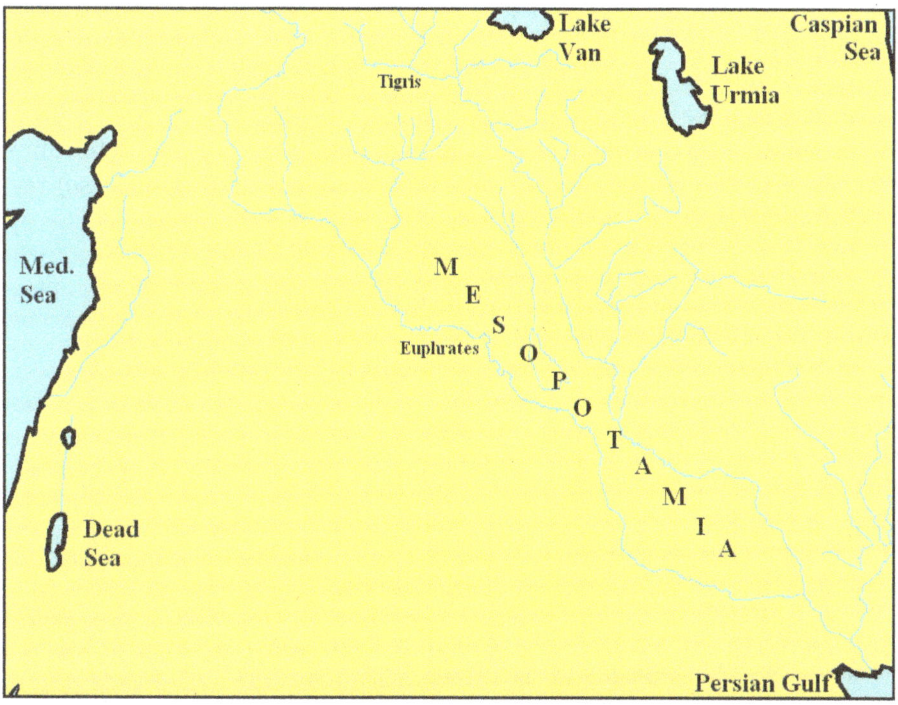

Figure 5.1: Location of Mesopotamia.

opening chapters of Genesis quickly locate the narrative at the confluence of the Tigris and Euphrates rivers,[1] near the head of the Persian Gulf.[2] Not until Noah's ark grounds within the mountainous regions of Ararat (southeast Turkey; near Lake Van in Figure 5.1) partway through the Flood does the geographical setting of the biblical narrative change.[3] And even then the setting quickly returns to Mesopotamia as Noah's descendants follow the Tigris and Euphrates rivers from their sources in the mountains of Ararat out onto the southern alluvial plains.[4] Thus it is fitting that a quest for the Flood within the data of secular disciplines should begin with South Mesopotamia.

Chronology of South Mesopotamia

To assist with archaeological orientation, Figure 5.2 shows a chronology of South Mesopotamia.[5] This chronology has been widely applied within the scientific literature. The periods are named after the archaeological sites in South Mesopotamia where pottery and other archaeological artifacts characteristic of that period were first discovered. This is predominantly an archaeological chronology, not a historical one. That is, it has been built up from archaeological data without the aid of historical documents (since no secular written materials are found prior to Late Uruk times). Archaeological stratigraphy has been used to determine the relative chronology, and this has been supplemented by (sparse) radiocarbon dates to help obtain an absolute chronology.

The Ubaid period is characterized by settled agricultural villages with abundant, decorated pottery and well-built multi-room houses. This characterization transforms into a fully urban society during the Uruk.

It would be very nice if Noah's Flood could be identified in the archaeology of South Mesopotamia by simply pointing to 3520 B.C., the biblical chronology date of the Flood, on this time chart. But it can't. The problem is that *accurate* secular chronologies of the distant past are

---

[1] Genesis 2:10–14.

[2] Gerald E. Aardsma, "The Location of Eden," *The Biblical Chronologist* 4.3 (May/June 1998): 1–5. www.BiblicalChronologist.org.

[3] Genesis 8:4.

[4] Genesis 11:2.

[5] Edith Porada, Donald P. Hansen, and Sally Dunham, "The Chronology of Mesopotamia, ca. 7000-1600 B.C.," in *Chronologies in Old World Archaeology*, volume 2 (Chicago: The University of Chicago Press, 1992), 94, 96, 98.

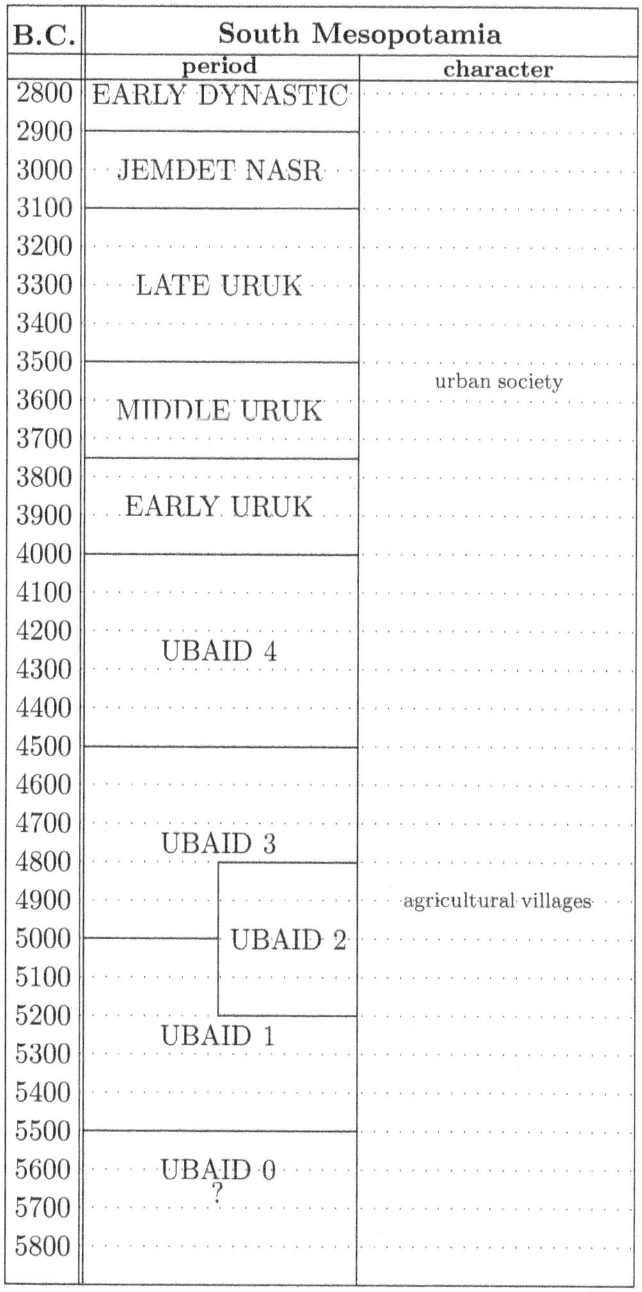

Figure 5.2: The COWA 1992 secular chronology of South Mesopotamia. Ubaid 2 overlaps Ubaid 1 and 3. The bottom boundary of Ubaid 0 has not yet been determined.

not all that easy to construct. Uncertainties of a few centuries are to be expected in archaeological chronologies such as this one when dealing with the remote millennium of interest to the Flood (i.e., the fourth millennium B.C.). The chronology shown in Figure 5.2 was constructed several decades ago, with very little assistance from radiocarbon dating, so it is certainly no exception to this general rule.

Chronological Reliability

That the chronology of South Mesopotamia is not yet settled can be seen by comparing Figure 5.2 with a corresponding chronology published in the *Cambridge Ancient History* two decades earlier.[6] There the Ubaid to Uruk boundary is found to be 500 years later (at 3500 B.C.) and the dawn of the Ubaid is well over a millennium later (at 4300 B.C.). Such large adjustments to this chronology over two decades of scholarly inquiry make it unlikely that the chronology shown in Figure 5.2 is the final answer.

And indeed, even more or less contemporaneously published chronologies of South Mesopotamia differ from that shown in Figure 5.2. For example, Kuhrt places the Late Uruk to Jemdet Nasr boundary a century earlier[7] as does Postgate,[8] and Postgate places the boundary between the Jemdet Nasr and the Early Dynastic periods also a century earlier.

These observations demonstrate that the chronology of South Mesopotamia at these early times is somewhat uncertain at present. This is likely to change over the next few decades as a result of radiocarbon dating having now clearly come of age. Improvements to the method since its inception over six decades ago—including tree-ring calibration, AMS measurement, and Bayesian analysis[9]—have summed to make radiocarbon dating ever more available, affordable, and powerful. But while a more accurate and reliable chronology of South Mesopotamia can be expected in the next few decades, for the present, inaccuracies of several

[6]I.E.S. Edwards, C.J. Gadd, and N.G.L. Hammond, ed. *The Cambridge Ancient History*, vol. 1, part 2. (Cambridge: Cambridge University Press, 1971), 997.

[7]Amélie Kuhrt, *The Ancient Near East*, Volume I (New York: Routledge, 1995), 22.

[8]J.N. Postgate, *Early Mesopotamia: Society and Economy at the Dawn of History* (New York: Routledge, 1994), 22.

[9]Alex Bayliss, "Rolling Out Revolution: Using Radiocarbon Dating in Archaeology," *Radiocarbon* 51.1 (2009): 123–147.

centuries in the Figure 5.2 chronology must be regarded as entirely possible.

As a result, Noah's Flood cannot be identified in the archaeology of South Mesopotamia by simply looking at the archaeology of the end of the Middle Uruk period, as Figure 5.2 might seem to imply. Rather, it is necessary to look into the archaeology of South Mesopotamia from at least the Early Uruk period through to the Jemdet Nasr period.

When this is done, the effort required is quickly rewarded by the discovery of what is being sought.

Archaeology of South Mesopotamia

An in-depth discussion of the archaeology of South Mesopotamia from the Early Uruk through to the Jemdet Nasr period, while highly interesting, is not needed for the present purpose. Rather, a direct route to data pertinent to Noah's Flood is preferred. This seems best provided by quoting briefly from a single article.

The article of interest, published in the science journal *Quaternary Research*, is coauthored by Michael Staubwasser of Hannover University, Germany, and Harvey Weiss, of Yale University. Their paper is not about Noah's Flood, of course. As noted at the outset of Chapter 1, there is presently a widespread consensus in academia that Noah's Flood is mythological only, and this consensus, together with traditionally mistaken chronological expectations, tends, rather strongly, to blind most researchers to any evidences of the Flood which may be present within their data.

The paper by Staubwasser and Weiss concerns itself only with past abrupt changes in climate and the impact these changes may have had on civilizations in the past. It identifies and discusses three abrupt climate change events at roughly 2200 B.C., 3200 B.C., and 6200 B.C. In all three cases, it finds that populations were reduced to a greater or lesser extent over large geographical areas.

Staubwasser and Weiss postulate widespread drought as the cause of these depopulation events in all three instances. However, the middle event, the one they date to 3200 B.C., uniquely exhibits the three signatures of Noah's Flood discussed in the previous chapter. They call this event "the late Uruk collapse." (Refer back to the Figure 5.2 chronology to understand this terminology.) They find it to be (1) a "hemispheric and possibly global"[10] event, (2) accompanied by widespread depopulation and (3) with very clear evidence, specifically in the archaeology of South Mesopotamia, of the inception of the institution of human government in an easily recognized, capital punishment-wielding form—that of monarchy.[11]

> Palace control of the southern Mesopotamian urban political economy appears to have emerged immediately after the late Uruk collapse following three thousand years of temple domination. The last temples of the late Uruk IV Eanna precinct were abandoned, replaced by terraces and light post and reed constructions. At the same time, archaeological excavation has retrieved the first palaces, administrative buildings distinguishable clearly from temples, at Jemdet Nasr, in mudbrick and then at Early Dynastic I period Uruk, in pisé. Synchronously, "council" rulership disappears from the proto-Sumerian lexicon and the title "king" is commonly documented.

Conclusion

Modern biblical chronology says the Flood happened in the middle of the fourth millennium B.C., not in the third millennium B.C. While nothing suitable to the Flood has ever been found anywhere in the third millennium B.C., when the focus is shifted to the middle of the fourth millennium B.C., using South Mesopotamia as the obvious pilot case, a suitable event—"the late Uruk collapse"—bearing three tell-tale signatures of the Flood immediately appears.

[10] Michael Staubwasser and Harvey Weiss, "Holocene Climate and Cultural Evolution in Late Prehistoric—Early Historic West Asia," *Quaternary Research* (2006), doi:10.1016/j.yqres.2006.09.001, 8.

[11] Michael Staubwasser and Harvey Weiss, "Holocene Climate and Cultural Evolution in Late Prehistoric—Early Historic West Asia," *Quaternary Research* (2006), doi:10.1016/j.yqres.2006.09.001, 9.

CHAPTER SIX

CHECK: ARCHAEOLOGICAL PERIODS

It is possible to carry out a quick check on the identification of "the Late Uruk collapse" with Noah's Flood without having to introduce anything new at this stage. This possibility presents itself because Genesis provides a history of South Mesopotamia subsequent to the Flood. The identification of "the Late Uruk collapse" with Noah's Flood may be checked by asking whether the archaeological periods embedded in the Figure 5.2 chronology of South Mesopotamia find any natural explanation relative to this most ancient of histories.

The answer is an unqualified yes.

The Check

In the previous chapter, it was found that the Flood brought the urban civilization of the Late Uruk to an end. Following the Flood, the biblical history recorded in Genesis 10 and 11 describes a significant, city-building society in South Mesopotamia, made up of the immediate descendants of Noah who settled in the land of Shinar (the archaeological *Sumer*) and who ultimately began construction of the Tower of Babel. The apparent unity of mankind up to Babel, and the Dispersion of mankind from Babel,[1] imply that this post-Flood, pre-Dispersion culture should be found only in South Mesopotamia. The Jemdet Nasr period

[1]Genesis 11:1–9.

(Figure 5.2), which appears from chronological charts to be found only in South Mesopotamia, immediately recommends itself for identification with this Tower-of-Babel culture.

The demise of the Tower-of-Babel / Jemdet Nasr culture, according to the biblical account, was due to the confusion of languages at Babel. The result was the Dispersion of mankind from Babel into surrounding lands. Genesis 10:25 records that the Dispersion happened in the days of Peleg (a name that means *division*). Genesis 11:10–16 say that Peleg was born about 100 years after the Flood, and Genesis 11:18–19 record that Peleg died when he was 239 years old.[2] Thus, the Dispersion must have occurred no sooner than about 100 years, and no later than about 340 years after the Flood. This requires that the Jemdet Nasr have a duration somewhere between 100 and 340 years. The 200-year duration for this period shown in the Figure 5.2 chronology satisfies this biblical requirement.

Following the Dispersion the biblical narrative naturally leads to an expectation of a geographically scattered emergence of state-controlled societies. The Early Dynastic period in South Mesopotamia and the parallel Early Dynastic period in Egypt, for example, fulfill this expectation.

Thus the secular and sacred records of post-Flood South Mesopotamia are found to harmonize readily in panoramic outline once the Flood has been situated at the end of the Late Uruk period, confirming identification of "the Late Uruk collapse" with Noah's Flood.

[2]Genesis ascribes very long lifespans to humans prior to the Flood (about 925 years on average), and the Bible records declining lifespans for about a thousand years following the Flood. Many have attempted to explain away these long lifespans because they contrast dramatically with present experience. There is, however, no sound scientific reason to do so. See, for example, Gerald E. Aardsma, "The Cause of Reduced Post-Flood Life Spans – Part III," *The Biblical Chronologist* 7.5 (September/October 2001): 1–5. www.BiblicalChronologist.org.

CHAPTER SEVEN

NOAH'S FLOOD IN PALESTINE

Now let us probe beyond the boundaries of South Mesopotamia.

Palestine (Figure 7.1) seems to be the next most obvious place to look for the Flood in secular scientific data. It has already been seen that,

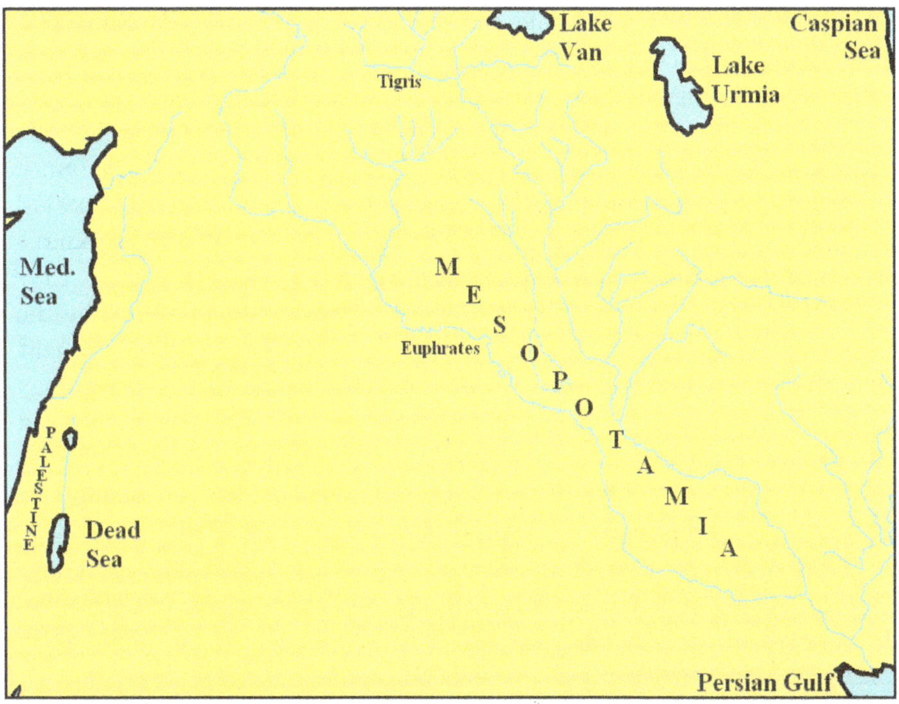

Figure 7.1: Location of Palestine. (Look to the lower left edge of the map.)

according to Genesis, the Flood extended far beyond local or regional boundaries. Palestine is relatively close to South Mesopotamia—only about one fortieth (2.5%) of the circumference of the earth away. It thus seems reasonable to expect the Flood to have extended to Palestine. In addition, Palestine has been subjected to extensive archaeological excavation for many decades now. In the quest for evidences of the Flood within the data of secular disciplines, all that is needed for a given region is a reasonably complete panoramic outline of the past, accompanied by a reasonably accurate (i.e., plus or minus a few centuries) chronology. These conditions are amply met in this region. Thus there is good reason to expect to find evidence of Noah's Flood in Palestine.

Chronology of Palestine

To help with archaeological orientation once again in this new region, Figure 7.2 shows a chronology of Palestine near 3500 B.C.[1] This chronology has been published fairly recently, so it may reasonably be expected to have better accuracy than the chronology of South Mesopotamia discussed previously (Figure 5.2).

The periods in this case are not named after sites. Rather, they were named, many decades ago, on the basis of evolutionary notions of the development of man and his tool assemblages. Chalcolithic (pronounce the "Ch" as a K) literally means "copper stone." The Chalcolithic period is followed by the Bronze, and then the Iron periods. While mankind's technological abilities have increased throughout history, just as they continue to do today, the simplistic evolutionary scheme imagined by the inventors of these period names has not been supported by subsequent archaeological research. These period names are retained by archaeology today only because of historical precedence, and should not be interpreted literally by the reader. Their utility lies in the fact that their divisions do signify cultural changes discernible to the archaeologist via changes in pottery styles, building styles, burial customs, and so forth.

This, once again, is primarily an archaeological chronology, not a historical one. Archaeological stratigraphy has been used to determine the relative chronology, and this has been supplemented by radiocarbon

[1]Chronological Tables, *The New Encyclopaedia of Archaeological Excavations in the Holy Land*, vol. 5 (Washington, DC: Biblical Archaeology Society, 2008), 2126.

B.C.	**Palestine**	
	period	**character**
2800		
2900	EARLY BRONZE II	
3000		
3100		
3200		urban
3300	EARLY BRONZE I	
3400		
3500		
3600		
3700		
3800		
3900		
4000		
4100		
4200		
4300		
4400		
4500		
4600		
4700	Chalcolithic	seminomadic
4800		
4900		
5000		
5100		
5200		
5300		
5400		
5500		
5600		
5700		
5800		

Figure 7.2: The NEAEHL 2008 secular chronology of Palestine.

dates to help obtain an absolute chronology.

Archaeology of Palestine

A complete loss of human population is, once again, the key signature of the Flood. This loss must take place at a period boundary in Palestine, just as it did in South Mesopotamia, because it is quite impossible to imagine that there could have been cultural continuity in Palestine from before the Flood to after the Flood. After the Flood it would have taken some time, at least decades, before Palestine could have been repopulated. The new population could hardly be expected to have done things in just the same way the pre-Flood population of Palestine had. For example, the new (post-Flood) population would surely have decorated its pottery in a way which was clearly different from the way the old (pre-Flood) population had decorated theirs. There would, in fact, have been no reason for the new population to have known very much about the old population which had lived there decades to centuries previously, and there would have been no reason to have imitated their culture even if they had known much about it.

Complete losses of human population over a wide region are, as already mentioned, relatively rare events. For Palestine, there appears to have been only one instance in which the whole land was completely depopulated. And this one instance falls exactly where anticipated—in the middle of the fourth millennium B.C.

Recall that the biblical chronology date for the Flood is 3520±21 B.C. The closest period boundary to this in Figure 7.2 is just 80 years away, at 3600 B.C. This boundary terminates the Chalcolithic. And secular archaeology in Palestine testifies to the fact that the Chalcolithic ended, uniquely, in a major depopulation event:[2]

> The impression is created of a sudden end to the period as a result of a catastrophe of some sort, ... Several suggestions have been offered to explain the disappearance of the people and culture of the Chalcolithic period. ... And where did all the know-how, sophistication, and originality of the Chalcolithic people in so many realms of creativity go? Those who

[2]Rivka Gonen, "The Chalcolithic Period," *The Archaeology of Ancient Israel*, ed. Amnon Ben-Tor (New Haven: Yale University Press, 1992), 79–80.

> followed them seem to have started from scratch, with the exception of some basic ceramic forms. All that had been attained during the Chalcolithic period disappeared, never to return, and the following generations never reached similar achievements, not even after hundreds and thousands of years.

To this evidence of depopulation may be added evidence in regard to human government. No evidence of palaces or government wielding capital punishment is found in Palestine during the Chalcolithic. In sharp contrast, the Early Bronze is characterized by city-states, each with its own ruler or king.

Conclusion

Modern biblical chronology says the Flood happened in the middle of the fourth millennium B.C. When the focus is placed on the middle of the fourth millennium B.C., evidences of the Flood, both in South Mesopotamia and in Palestine, are immediately found.

In Palestine, the offset between the biblical chronology date of the Flood (i.e., 3520±21 B.C.) and a secular date for the expected archaeologically observed depopulation event, using a modern secular chronology of Palestine, (i.e., 3600 B.C.), is found to be just eighty years. This small difference is well within dating uncertainties at such a remote date. Thus the two events may reasonably be regarded as synchronous, which is the same as saying that they are one and the same event. Noah's Flood wiped out the population of Palestine, bringing its Chalcolithic culture to an abrupt end.

The evidences of the Flood, found at the expected time in Palestine and Mesopotamia, show that the Flood was a real historical event. It was catastrophic to mid-fourth-millennium civilizations in the Middle East, but it left the remains of those civilizations relatively undisturbed and accessible to modern archaeology. Thus, while the surface has barely begun to be scratched in regard to the true historical nature of Noah's Flood, it is already apparent that the Flood is neither a myth nor an earth-shattering tectonic cataclysm.

CHAPTER EIGHT

CHECK: FILLING OF THE DEAD SEA DEPRESSION

It is possible at this stage to carry out another quick check. The check is possible this time because of some unique geomorphology in Palestine—specifically, the Dead Sea depression.

The thinking behind this check is simple. If the Chalcolithic peoples in Palestine were swept away by Noah's Flood, then the Dead Sea depression must have been filled up with Flood water at the end of the Chalcolithic (i.e., in the middle of the fourth millennium B.C.). Is there any evidence for a filling of the Dead Sea depression in the middle of the fourth millennium B.C. within the secular scientific data?

Indeed, there is.

The Check

The Dead Sea depression may be compared to a giant bathtub with no drain. Water normally enters this "bathtub" from precipitation within the Dead Sea catchment area. It can leave only by evaporation.

The Dead Sea itself is small relative to the size of the Dead Sea depression. The surface of the Dead Sea today is somewhat more than 400 meters below sea level. The Dead Sea depression can be filled to 60.5 meters *above* sea level before it will begin to spill over.[1] The volume of a

[1] Tina M. Niemi, Zvi Ben-Avraham and Joel Gat, eds. *The Dead Sea: the Lake and its Setting*, (New York: Oxford University Press, 1997), 19 (Figure 2-6).

lake filling the Dead Sea depression would be twelve times the volume of the Dead Sea at present, and the surface area would be over nine times larger than the surface area of the Dead Sea at present.

The Data

In 1991, in the science journal *The Holocene*, Frumkin *et al.* published a graph of past fluctuations of Dead Sea level.[2] The solid line in Figure 8.1 shows their basic graph. (I have converted the published time axis from "^{14}C years BP" to calendar years for ease of use in the current context.[3])

This graph immediately shows that, while the surface of the Dead Sea has been more than 375 meters below mean sea level for most of the past eight thousand years, it was significantly above this normal elevation for roughly a thousand years, *beginning in the middle of the fourth millennium B.C.* Said another way, the researchers found that the Dead Sea was at an unusually high stand only once in the past seven and a half thousand years, and this one instance coincides closely with the modern biblical chronology date of the Flood.

There are several questions which must be addressed with this graph before it can be concluded that this second check is satisfied. First, why does the rising edge of the (solid line) high-stand peak not coincide exactly with the Flood at 3520 B.C.? Second, if this high stand was due to the Flood, why does the solid line curve reach to only -300 meters? While this is remarkably high, it is still very much less than for a full Dead Sea depression, which today is +60.5 meters. And third, is it reasonable for it to take a thousand years to get the Dead Sea back to normal (i.e., below -375 meters) once the Dead Sea depression has been filled with Flood water? But before these questions are tackled, notice that this graph might easily have falsified the claim that Noah's Flood

---

[2]A. Frumkin, M. Magaritz, I. Carmi, and I. Zak, "The Holocene Climatic Record of the Salt Caves of Mount Sedom," *The Holocene* 1.3 (1991): 191–200.

[3]This conversion was made using CALIB 4.0. (See M. Stuiver and P. J. Reimer, "Extended ^{14}C Data Base and Revised CALIB 3.0 ^{14}C Age Calibration Program," *Radiocarbon* 35.1 (1993): 215–230.) I first reproduced the elevation versus ^{14}C years BP curve of Frumkin *et al.* using piecewise quadratic interpolation, then calibrated the time coordinate of all of the interpolation points to show the elevation curve in calendar years. I would have preferred to redraw the elevation curve from scratch having first calibrated the radiocarbon dates of the organic samples upon which the Frumkin *et al.* curve is based, but this was not possible because I was unable to find a relationship between passage width ratio and elevation in the Frumkin *et al.* paper.

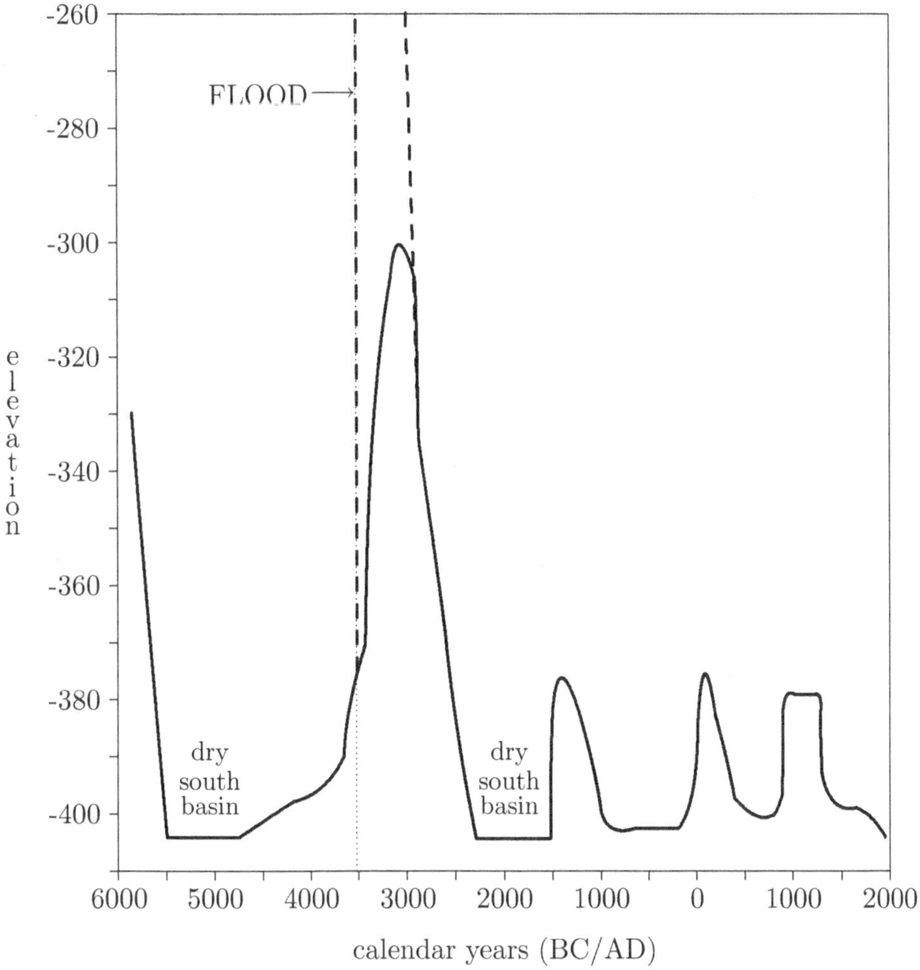

Figure 8.1: Elevation data from Frumkin *et al.* (*The Holocene*, 1.3 (1991): 191–200) for the surface of the Dead Sea in meters above mean sea level versus calendar year (solid curve). I have added two dashed lines that, when extended, intersect at +60.5 meters, far above the graphed region, yielding a peak approximating the probable actual elevation of the Dead Sea surface due to filling of the Dead Sea depression by Noah's Flood.

caused the termination of Chalcolithic civilization in Palestine. All that was required was that the Dead Sea surface should stay below -375 meters the whole time. But it didn't. Instead it went in the opposite direction, immediately showing an unusual high stand in near coincidence with the Flood. This seems more than happenstance, even if some details remain to be clarified.

Why Not Exact Coincidence?

The failure to find exact coincidence between the leading edge of the high-stand peak and the date of the Flood is not surprising. It results from the fact that building accurate chronologies of the distant past is not a trivial exercise—as has previously been pointed out.

Figure 8.1 is a chronology of the Dead Sea elevation. The time axis for this chronology is provided by radiocarbon dates on several dozen wood samples. (More about this below.) The radiocarbon dates in this instance are not terribly precise, and this introduces significant uncertainties into the time axis of the Figure 8.1 chronology. These uncertainties make the placement of the leading edge of the high-stand peak uncertain within plus or minus several centuries.

Radiocarbon is a means of measuring elapsed time—the time from the death of a once-living thing to the time of measurement of its residual radiocarbon. Experimental error cannot be reduced to zero for any physical measurement; the measurement of elapsed time using radiocarbon is no exception. As a result of measurement error, radiocarbon does not provide a single unique date for a sample. Rather it provides a range of dates in which the true date of the sample likely lies.

In the present case, it is the leading edge of the high-stand peak which is of interest. It is expected to coincide with the Flood at roughly 3520 B.C. Consider the two radiocarbon dates on either side of this leading edge. These two dates obviously most strongly influenced where the researchers drew the leading edge. One date ranges from 4220 B.C. to 3363 B.C.[4] The other ranges from 3628 B.C. to 3020 B.C. Both of these date ranges include the date of the Flood, 3520 B.C. It is thus clear that it is possible to draw the high-stand peak with its leading edge in exact coincidence with the Flood.

Said simply, exact coincidence between the date of the Flood and the

[4] CALIB 4.0; 2σ.

leading edge of the high-stand peak drawn by the original researchers is not observed because the relatively poor precision of the radiocarbon dates upon which the Figure 8.1 time axis is based leads to an intrinsic uncertainty of plus or minus three hundred to four hundred years in the Figure 8.1 chronology of the Dead Sea elevation back at the time of the Flood.

Why Only -300 Meters?

To answer this second question, yet more of the process used to construct the solid line curve must be understood.

The solid line curve results from measurements made on caves in a mountain which borders the Dead Sea on its southwestern shore. The mountain is called Mount Sedom today. It is a mountain composed of salt, with a cap of rock. The height of this mountain sets an upper limit on the past elevation of the surface of the Dead Sea which can be detected using this measurement method. This limit appears to be roughly -300 meters.

Rain on top of the mountain has, for years, worked its way into fissures of the rock cap and dissolved caves through the underlying salt body of the mountain. These water-dissolved conduits empty into the south basin of the Dead Sea. The width of the caves and their height in the mountain can be used to reconstruct the elevation of the Dead Sea surface when they were in use. When the level of the Dead Sea changes relative to the mountain, a new channel will be rapidly cut by the fresh water runoff through the mountain. The date when a cave was in use can be determined by radiocarbon dating organic debris, such as small twigs, found in gravels washed off the mountain and into the caves. The radiocarbon dates discussed above come from this organic debris.

Mount Sedom is 11 kilometers long and 1.5 kilometers across. Its cap of rock has a maximum thickness of about 40 meters. The underlying salt rises to a maximum of only about -200 meters below mean sea level today.

The mountain is rising today by at least 3.5 millimeters per year, and is assumed to have been rising at roughly the same rate for the entire duration of the Figure 8.1 graph. This means that back at the time of the Flood the salt of the mountain rose to a maximum elevation of only -220 meters below mean sea level.

This sets a fundamental limitation on this method of determining the level of the Dead Sea in the past. When the Dead Sea surface is higher than the mountain, no caves will be formed in the mountain because the mountain will be entirely under the salt water of the Dead Sea. The measurement method is thus limited by the height of the mountain. Were the Dead Sea depression to be filled to the brim, this method of measuring Dead Sea elevation would be unable to detect such a filling. The highest level it can possibly detect today is -200 meters. And since water requires some gradient to flow in conduits through the mountain, it may be anticipated that the actual practical limit for this measurement method is significantly less than this. For example, a fairly shallow, one-in-twenty slope, for a conduit through a 1.5 kilometers-wide salt mountain, will result in the water emerging from the mountain 75 meters lower down than it entered the mountain. The oldest dated cave actually found by Frumkin *et al.* lies directly beneath the rock cap at an elevation of just -290 m. Because the mountain is rising, this cave would have been tens of meters lower when it was formed. Thus it appears that, for a filling of the Dead Sea depression back at the time of the Flood, something close to -300 meters is the actual practical limit on the maximum elevation measurable by this method.

This seems to be the fundamental reason why the solid curve reaches to only -300 meters. Had Mount Sedom been another 100 meters higher back at the time of the Flood, there would, no doubt, have been data points from caves at higher elevations, showing that the high-stand peak reached well above -300 meters.

I have drawn dashed lines in Figure 8.1 to show approximately what might be expected for the Dead Sea depression filling to the brim at the time of the Flood. These lines intersect far above the graph area, at +60.5 meters. Mount Sedom would then have been far under water, and the salt part of the mountain would have remained under water until the surface of the Dead Sea had fallen below at least -220 meters. During that time no caves would have been formed, and no wood would have been deposited in the caves.

Eventually, the mountain would have emerged from the receding (evaporating) waters. Twigs and other organic debris which had been floating in the lake would have become stranded on the mountain as the lake's surface dropped ever lower relative to the mountain. Eventually, some of these organics would have been washed, by rainwater, into the caves.

Putting this all together, the overall solid curve is a very rough interpolation of roughly thirty data points. The curve has been drawn by eye through the data points by the original researchers. Each point is made up of a radiocarbon date on some wood sample found within the cave system and the corresponding cave geometry. The data points show considerable scatter for both elevation and time. The peak of interest, the high stand peak, contains five scattered data points, the highest of which has been used to fix the height of the peak. Because the measurement method shuts down for very high surface levels of the Dead Sea, the dashed curve I have drawn in Figure 8.1, corresponding to a complete filling of the Dead Sea depression, is just as valid an interpolation of the actual data points as the solid curve supplied by the original researchers.

A Millennium to Dry?

But is a thousand years a reasonable time for the lake to take to dry down again after the Flood? Here, too, the answer appears to be yes.

The actual time required to evaporate the newly-filled lake back down to normal levels is actually too complex to try to calculate quantitatively. It depends on several factors, such as salinity of the lake waters (which will be increasing as the water is evaporated away), temperature (which will rise on average as the lake surface lowers), winds (which will change relative to the surface of the lake as the surface drops lower in the basin), and precipitation in the Dead Sea catchment basin (which may be expected to be impacted by the extent of the lake itself).

Rather than try to calculate any of this, it seems better simply to observe the recent behavior of the Dead Sea itself. Its supply of water (mainly via the Jordan River) has been severely depleted in recent years for use in irrigation and drinking water. This has disturbed the recent historical equilibrium level of the Dead Sea and has resulted in a loss of elevation of the Dead Sea surface of roughly one meter per year.

Using this as a ball-park figure for rate of recession of the surface of the Sea when much out of equilibrium yields 435 years as the time required to restore the Dead Sea to normal levels following a complete filling of the basin.

This is very approximate, of course, as discussed above. But it is adequate to show that the Dead Sea would not have returned to normal levels in a year or a decade or even a century following the Flood. A millennium seems indeed to be a reasonable expectation.

Conclusion

This second check is thus satisfied. Secular science does indeed find an unusual high stand of the Dead Sea in the middle of the fourth millennium B.C.

Chapter Nine

Objection: Dead Sea Sediments and Old Shorelines

Before moving on to additional data showing Noah's Flood in the middle of the fourth millennium B.C., it is necessary to pause and deal with two objections. The first objection, dealt with in the present chapter, is that, in contradiction to the findings of the previous chapter, Dead Sea sediments and missing old shorelines show that the Dead Sea was not filled by Noah's Flood. The second objection, dealt with in the next chapter, is that some salt-covered land snail shells show the same thing.

General Background

Geologist David Neev of the Geological Society of Israel and oceanographer K. O. Emery of the Woods Hole Oceanographic Institution in Massachusetts coauthored a book, *The Destruction of Sodom, Gomorrah, and Jericho,* published in 1995, purporting to give geological, climatological, and archaeological background to the biblical accounts of the destructions at Sodom and Gomorrah and at Jericho.[1] If you wonder why a geologist and an oceanographer should undertake such a book—seeming, as it does, somewhat removed from either author's field of expertise—

[1]David Neev and K. O. Emery, *The Destruction of Sodom, Gomorrah, and Jericho,* (New York: Oxford University Press, 1995).

the authors explain as follows (square brackets indicate amplification by me):[2]

> Our studies of the Dead Sea region [in which Sodom, Gomorrah, and Jericho are located] began about 35 years ago as an effort to learn how salt (sodium chloride) is deposited in lakes and seas—an initiative impelled only by scientific curiosity. ...Samples of bottom sediments [from the Dead Sea] revealed prior history of rocksalt deposition and, thus, of climate during several thousand years. This information supplemented inferences derived by archaeologists from remains of settlements and changes of culture. Other relevant geological knowledge for the region comes from the spatial and temporal distribution of earthquakes and volcanic activity that had evolved from vague ancient tradition to precise modern instrumentation. Assembly of these various kinds of information gave such promise for solving the fates of ancient Sodom and Gomorrah that we were encouraged to extend the effort to learn about the destruction of Jericho at a somewhat later date and to the legend of Noah's Flood at a much earlier date.

My assessment of Neev and Emery's book is that: 1. it is a very handy primer on much of scientific interest to the Dead Sea, and 2. that it fails in its primary purpose of illuminating the destructions in question.

The second point in my assessment above stems from the fact that the authors' chronologies of the destructions in question lack both sufficient accuracy and the necessary precision for them to address their subject meaningfully. But that is of little interest in the present context.

What is of interest here is the fact that, as the foregoing quote mentions, they included some treatment of Noah's Flood in their book, despite its exclusion from their title.

Their basic conclusion in regard to the Flood is negative. They conclude that the biblical narrative of Noah's Flood is, at best, badly exaggerated. In particular, the biblical claim that the Flood covered high mountains, they assure us, cannot be true.

---

[2]David Neev and K. O. Emery, *The Destruction of Sodom, Gomorrah, and Jericho*, (New York: Oxford University Press, 1995), vi–vii.

In one sense, their conclusion is a little surprising. It is a little surprising because their combined expertise justifying their book centers around the Dead Sea, as the quote above has just explained. And as has just been seen in Chapter 8, the Dead Sea offers some unique evidence in support of the historical reality of Noah's Flood.

Yet in another (if somewhat disappointing) sense, their conclusion is not surprising. As mentioned in Chapter 1—it bears repeating—the consensus view in academia today is that Noah's Flood is mythological only, and this consensus, together with traditionally mistaken chronological expectations, tends, rather strongly, to blind most researchers to any evidences of the Flood which may be present within their data.

As a case in point: Neev and Emery are aware that the Chalcolithic civilization in Palestine simply disappeared. They write:[3]

> Ghassulians [Chalcolithic peoples] appear to have arrived in Canaan from somewhere else in the Mideast after a hiatus at the end of the Pottery Neolithic age. Later they disappeared without indications of conquest or destruction of their settlements; they simply vanished.

They also show unusual familiarity with the story of the Flood, both as it appears in the Bible and in extra-biblical sources, and they are thus fully aware that this unusual depopulation event may well have been synchronous with the Flood:[4]

> . . . Wet Subphase II-a could have happened sometime in Chalcolithic about 6500 to 6000 B.P. The latter dates are near enough to the time of the epic flood [Noah's Flood] described in *Gilgamesh* . . . According to the story, the ship survived the flood about 6000 B.P. . . . .

So Neev and Emery are aware:

1. that the Bible says the Flood nearly exterminated humanity,

2. that an entire civilization "vanished" from Palestine,

[3]David Neev and K. O. Emery, *The Destruction of Sodom, Gomorrah, and Jericho*, (New York: Oxford University Press, 1995), 101.

[4]David Neev and K. O. Emery, *The Destruction of Sodom, Gomorrah, and Jericho*, (New York: Oxford University Press, 1995), 119.

3. that there is no obvious archaeological explanation for the disappearance of this civilization, and finally

4. that this civilization may well have vanished at the same time the Flood happened.

But the penny never drops. Neev and Emery are in full possession of the most obvious signature of Noah's Flood, yet this signature appears to escape their notice entirely. Rather, on the basis of far more equivocal data, which I will now discuss, they conclude, rather strongly, that the biblical version of Noah's Flood is, at best, seriously overstated.

The Objection

Neev and Emery's basic claim is that, in contradiction to the findings in Chapter 8, the Dead Sea was never filled to the brim anytime during the past 11,500 years:[5]

> No evidence has been found for a world-encircling flood during the Holocene [roughly the last 11,500 years] either from [Dead] sea levels high enough to fill the graben [Dead Sea depression] and leave high shorelines or from marked dilution of brines recorded in sediments from oceanic water deposited on the floor of the [Dead] sea. ... There is no recognized evidence for dilution of brine in the graben that would be expected if large enough volumes of water were added to the ocean to allow [from the Gilgamesh Epic] Utnapishtim's ship to be stranded on Mount Nisir, perhaps 2,700 m above sea level, or later for Noah's ark on Mount Ararat (5,137 m) or its flank. ...
>
> Absence of evidence within sediments means that these floods, no matter how enormous, were not high enough to have raised ocean level sufficiently to submerge the sills that separate the Dead Sea depression from the ocean so that flow could pass through the Jezreel or Arava Valleys to the Dead Sea graben.

[5]David Neev and K. O. Emery, *The Destruction of Sodom, Gomorrah, and Jericho*, (New York: Oxford University Press, 1995), 120.

To clarify, Neev and Emery are saying that if Noah's Flood had been deep enough to perch the ark on a high mountain, then it would also have been deep enough to overtop the sill separating the Dead Sea from the Mediterranean to the east and the sill separating the Dead Sea from the Gulf of Aqaba to the south and fill up the Dead Sea depression with ocean water. And if this had happened, the high resultant Dead Sea levels would have left high shorelines, and the influx of ocean water would have diluted the Dead Sea brine, and this dilution would have been recorded in the sediments on the bottom of the Dead Sea. Their claim is that the Dead Sea was not filled by Noah's Flood. They feel they know this on the basis of two assertions: (1) there are no high shorelines, and (2) there is no evidence in Dead Sea sediments of dilution of brine.

Both of these assertions are seriously in need of some amplification by the authors. It is as if the authors have expected general agreement with their conclusion—that the Bible is badly mistaken in regard to Noah's Flood—and as a result, have not bothered to do the sort of work which is necessary to build a sound case.

Old Shorelines

Lakes produce shorelines, or beaches, through the normal action of waves. If a lake's surface changes from a high level to a low level, a new shoreline will be formed at the low level, and the old shoreline will be left as a relict, high and dry above.

Should the filling of the Dead Sea depression by the Flood be expected to have produced old shorelines?

Very probably. I cannot say "certainly" because it takes time for a beach to be formed by wave action, and, as was seen in Chapter 8, it appears that the level of the Dead Sea decreased fairly rapidly after the Flood. Recall that it seems to have taken only about 1000 years for the Dead Sea to have gotten back down to normal levels (below -375 meters) from topped up full level (roughly +60 meters). This means that on average the Dead Sea surface elevation was decreasing by 0.4 meters (roughly one and a half feet) per year. On a 45 degree slope, the beach would have receded about 2 feet per year on average.

Would this rate have been slow enough to produce old shorelines? While it cannot be concluded with certainty, it nevertheless seems quite probable that somewhere between brim full and normal level the altitude of the surface of the Dead Sea would have remained stable long enough, and wave action would have been severe enough, to produce at least one old shoreline in at least one location somewhere around the perimeter of the Dead Sea.

It seems reasonable that the surface of the Dead Sea would have dropped faster when the Dead Sea depression was full, than when it was near to normal levels. This is because high surface elevations mean greater exposure to both sun and wind, aiding evaporation, and salinity would have increased as water evaporated off and the surface lowered, impeding further evaporation. This means that the shoreline would have been receding at its fastest rate initially. So there is probably a much better chance of an old shoreline having been left by the Flood lower down than higher up in the Dead Sea depression.

Now it must be asked whether an old shoreline due to a Flood filling of the Dead Sea depression five and a half thousand years ago should be identifiable at present? This is a much more difficult problem.

There is, of course, the matter of preservation. Whatever shorelines were formed by the Flood-filled Dead Sea would need to be able to survive obliteration by five and a half thousand years of natural erosion to be observable today. Obviously, the longer the surface of the Dead Sea could be maintained at a given elevation, the better the chances of carving a shoreline that would persist. But because the Flood filling was a one-time event, the subsequent Dead Sea would be expected to be characterized by a steadily decreasing surface altitude, not prolonged stasis. Thus, preservation of old shorelines due to the Flood filling may reasonably be expected to be a problem.

But there is a yet greater problem. This problem arises mainly because the Dead Sea depression has been filled with water more often than just at the time of the Flood. Specifically, it is common knowledge that the Dead Sea depression was occupied by a large saltwater lake called "Lake Lisan" up until 10 or 12 thousand years ago. Lake Lisan is believed to have occupied the Dead Sea depression from about 70 thousand years ago, with fluctuating surface levels 100 to 250 meters higher than today. Obviously, Lake Lisan had opportunity to leave old shorelines too. In fact, it had some 60 times more opportunity to leave old shorelines than the Flood filling did because it existed for 60,000 years while the Flood filling was gone in just 1000 years.

This creates an obvious problem. How is it possible to tell whether an old shoreline is due to Lake Lisan or due to the Flood?

The obvious answer to this question is that it will be necessary to date old shorelines using some physical dating method. Perhaps the old shoreline could be dated by applying radiocarbon to organic debris (e.g., driftwood) found buried in the old beach. The date would then tell whether that old shoreline was due to Lake Lisan or to the Flood.

Now let me bring this all home to Neev and Emery's assertion.

Neev and Emery say the Flood left no old shorelines around the Dead Sea. I challenge them to prove it. To do so, they will first need to catalog every old shoreline anywhere within the Dead Sea depression. This will be mandatory because their claim depends upon the validity of a universal negative: no old shoreline due to the Flood exists. They will then need to apply some physical dating method to each individual one of these old shorelines to tell whether it is due to Lake Lisan, the Flood, or indeed some other high surface level entirely.

This necessary proof would be fairly easy if there were, in fact, no old shorelines of any sort to be found. (But this, unfortunately, would defeat the argument entirely. For if Lake Lisan had failed to produce an old shoreline in 60 thousand years, why should the Flood filling be expected to produce old shorelines in just one thousand years?) Otherwise, the proof is very difficult and probably a practical impossibility.

In point of fact, there are very definitely old shorelines surrounding the Dead Sea. Neev and Emery show two examples of sets of old shorelines in photographs in their book. The first, their Figure 2.11, has in the caption:[6]

> Horizontal lines beyond it [the valley of Nahal Zohar] are lake terraces or shorelines that reach to -180 m m.s.l.

The second, their Figure 2.16, has in the caption:[7]

> Sequence of small terraces north of Mount Sedom formed at successive elevations due to changing levels of Lisan Lake and the [Dead] sea.

---

[6]David Neev and K. O. Emery, *The Destruction of Sodom, Gomorrah, and Jericho*, (New York: Oxford University Press, 1995), 28.

[7]David Neev and K. O. Emery, *The Destruction of Sodom, Gomorrah, and Jericho*, (New York: Oxford University Press, 1995), 35.

To be accurate, Neev and Emery should probably have said something like, "No old shorelines in the Dead Sea depression have ever been demonstrated to be due to Noah's Flood, but then nobody has ever searched there for old shorelines due to Noah's Flood because nobody believes the Flood happened the way the Bible says anymore, and all old shorelines which do exist have simply been assumed to be due to Lake Lisan, so whether there are any old shorelines there due to Noah's Flood is really unknown at this time."

In any event, Neev and Emery's assertion that there are no old shorelines due to the Flood is unproven and probably impossible to prove. Let us move on.

Dilution of Brine

Unfortunately, Neev and Emery's assertion regarding dilution of brine fares even worse.

The "water" filling the Dead Sea today is, in fact, a brine—a strong saline solution. Salts are continuously brought into the Dead Sea by the Jordan and other rivers. The only way water can get out of the Dead Sea at present is by evaporation. As the water evaporates, the salts are left behind. Hence the brine. When the brine gets too saline, it begins to precipitate rock salt, which collects on the floor of the Dead Sea as a sediment.

Wetter conditions give rise to marl rather than rock salt as the sediment. Marl, according to *Webster's Ninth New Collegiate Dictionary,* is "a loose or crumbling earthy deposit (as of sand, silt, or clay) that contains a substantial amount of calcium carbonate and is used esp. as a fertilizer for soils deficient in lime."

The Dead Sea has two basins, a deep north basin, and a more shallow south basin. Almost all of the coring of sediments which has been done at the Dead Sea reported by Neev and Emery has been done in the south basin. The sediments of the south basin tend to be more interesting than those of the deep north basin. The shallow south basin is much more prone to drying out completely than is the deep north basin. When the south basin dries out (goes below the critical level of about -400 meters) it precipitates rock salt to its floor. Thus a layer of rock salt in the sediments of the south basin indicates that times were dry; marl indicates more moist times.

filling the Dead Sea depression with ocean water at the Flood would dilute the Dead Sea brine. Dilute brine will not precipitate rock salt—notice that the oceans do not precipitate rock salt—so marl will be the initial sediment following the Flood. If the south basin had been dry prior to the Flood, there would be a sudden change from rock salt sediment to marl sediment at the time of the Flood. This would be quite noticeable in the sediment column today. If, on the other hand, times had been relatively moist prior to the Flood, then filling the Dead Sea depression with ocean water would simply result in more marl sediment being added to the existing marl sediment. This would be much less noticeable and might, in fact, be indistinguishable today from the pre-Flood marl.

The best information I am aware of at present in regard to moist and dry intervals for the Dead Sea is the data by Frumkin *et al.* already introduced in Chapter 8. These data, shown in Figure 8.1 (page 63), show that conditions were relatively moist prior to the Flood. So, in evident contradiction to Neev and Emery's assertion, the Flood is expected to have made no change in the type of sediment due to dilution of the brine.

It is possible to go further.

While dilution of the Dead Sea brine by ocean water at the time of the Flood is not expected to have brought about a change in the type of sediment being deposited, the influx of ocean water at the time of the Flood would have necessitated a deposition of rock salt in the centuries following the Flood.

Ocean water contains roughly 35 grams of salts per liter, most of which is NaCl (table salt). Rock salt precipitates from warm water when the concentration of NaCl exceeds about 370 grams per liter. It was mentioned previously (Chapter 8, page 62) that the volume of the Dead Sea depression is about 12 times the present volume of the Dead Sea. So, by the time the volume of ocean water in the Flood-filled Dead Sea depression had evaporated down to the present Dead Sea volume, the concentration of the brine would have been at least $(35 \times 12 =)$ 420 grams per liter. Since this is greater than the 370 grams per liter at which rock salt precipitates from warm water, it is clear that the sediment type would have changed to rock salt prior to the complete desiccation of the south basin following the Flood.

This change of sediment type did take place.[8] But this change of

[8]A. Frumkin, M. Magaritz, I. Carmi, and I. Zak, "The Holocene Climatic Record of the Salt Caves of Mount Sedom," *The Holocene* 1.3 (1991): Figure 8, page 197.

sediment type is also expected in the case of the Dead Sea high stand at the end of the Chalcolithic (Chapter 8) *not* being due to the Flood.

Putting this all together, it is found that, on the basis of sediment type, it is impossible to distinguish the case of the Flood having filled the Dead Sea depression from the case of it not having filled the Dead Sea depression. Thus, Neev and Emery's conclusion: "Absence of evidence within sediments means that these floods, no matter how enormous, were not high enough to have raised ocean level sufficiently to submerge the sills that separate the Dead Sea depression from the ocean so that flow could pass through the Jezreel or Arava Valleys to the Dead Sea graben" is not substantiated.

Conclusion

Neev and Emery's claim that Dead Sea sediments and missing old shorelines show that Noah's Flood failed to fill the Dead Sea depression appears to be vacuous.

Chapter Ten

Objection: Salt-Covered Snail Shells

A refuse heap (or midden) made by rodents was found a few decades ago under a large boulder on the scree slope down to the Dead Sea behind Qumran. The midden was partially encrusted with salt. It contained shells of land snails covered by crystallized salt.

The salt crust made it immediately clear that this location had once been in close proximity to the shoreline of the Dead Sea. The salt crust would have resulted from droplets of brine sprayed from Dead Sea waves onto the shells. The aridity of the region surrounding the Dead Sea guaranteed rapid evaporation of the water from the sprayed coating, leaving behind the salt.

The midden was located at -280 meters elevation. This is 120 meters higher than the present surface of the Dead Sea—about the height of a skyscraper of forty stories. The Dead Sea had obviously been much higher at some point in the past.

The land snails had served as food for the rodents. The emptied shells had then been discarded in the midden. At some subsequent time, the Dead Sea level had risen sufficiently to bring its shoreline within reach of the midden.

After encrusting the midden, the shoreline had receded. The boulder had sheltered the midden from (scarce) rainfall in the area, preserving the salt crust from this old shoreline into modern times.

Researchers Goodfriend *et al.* describe the midden and measurements made on it.[1] They date the old shoreline responsible for the salt crust to the "end of the Neolithic," which they place roughly 5500 B.C. They claim:[2]

> Water level may have been at that elevation for several years, depositing successive sublayers [of salt], and then went down, never again to rise above that level (since this would have immediately dissolved and washed away the salt).

This yields the objection: Salt-covered land snail shells show that the Dead Sea depression was not filled at the time of the Flood. If the Dead Sea depression had been filled by the Flood 3520±21 B.C., then that filling would have washed away the much earlier 5500 B.C. salt from the shells.

Mistaken Logic

The logic leading to this objection is faulty. The date of the salt crust is, in fact, unknown.

Goodfriend *et al.* obtained the maximum age of the salt crust by radiocarbon dating the land snail shells. They found a radiocarbon date range for the shells of 6660±400 yr BP. This yields a maximum age of roughly 7100 radiocarbon years BP (add 400 to 6660 and round to the nearest one hundred). When calibrated using tree rings, this becomes roughly 5950 B.C.[3]

The logic to this point is fine. The midden dates to roughly two thousand years before the Flood and probably not earlier than 5950 B.C.

But what is the date of the salt crust and corresponding old shoreline? In principle, the crust could have been formed by a high enough Dead Sea level at any subsequent time.

---

[1] Glenn A. Goodfriend, Mordeckai Magaritz, and Israel Carmi, "A high stand of the Dead Sea at the end of the Neolithic Period: Paleoclimatic and archeological implications" *Climatic Change* 9 (1986): 349–356.

[2] Glenn A. Goodfriend, Mordeckai Magaritz, and Israel Carmi, "A high stand of the Dead Sea at the end of the Neolithic Period: Paleoclimatic and archeological implications" *Climatic Change* 9 (1986): 352.

[3] Minze Stuiver and Bernd Becker, "High-Precision Decadal Calibration of the Radiocarbon Time Scale, AD 1950–6000 BC," *Radiocarbon* 35.1 (1993): 56.

Goodfriend *et al.* say:[4]

> The pure-white appearance of the salt suggests that this [deposition of salt] happened sometime shortly after the deposition of the shells, before loessial [wind-borne] sediments had time to accumulate.

But this is obviously inadequate. One can reasonably postulate, for example, that a much later rise of the Dead Sea surface inundated the midden, washing away not only any earlier salt crust but also whatever loessial sediments may have accumulated by that time, leaving the shells exposed and clean, ready for a fresh deposit of salt when the Dead Sea surface receded once again.

Goodfriend *et al.* overlook this possibility and press on to obtain a minimum date for the salt crust by using radiocarbon dates on archaeological remains bordering the Dead Sea.[5]

> Radiocarbon dates from the archaeological site of Teleillat Ghassul north of the Dead Sea greatly restrict the time when the salt deposition could have occurred. This Chalcolithic settlement with a base at -295 m elevation has a series of radiocarbon dates ranging from 6550±160 yr BP to 5500±110 yr BP. Thus by 6400 yr BP [roughly 5330 B.C.[6]] or earlier, the level of the Dead Sea must have fallen to well below the level of the salt deposit at Qumran.

The logic here is obviously flawed. There is nothing whatsoever to exclude the possibility that the Flood submerged both the archaeological site of Teleillat Ghassul and the rodent midden at Qumran by a complete filling of the Dead Sea depression 3520±21 B.C., and that it is this filling which left behind the presently observed salt crust on the shells as the Dead Sea surface receded by evaporation in subsequent centuries.

---

[4]Glenn A. Goodfriend, Mordeckai Magaritz, and Israel Carmi, "A high stand of the Dead Sea at the end of the Neolithic Period: Paleoclimatic and archeological implications" *Climatic Change* 9 (1986): 352.

[5]Glenn A. Goodfriend, Mordeckai Magaritz, and Israel Carmi, "A high stand of the Dead Sea at the end of the Neolithic Period: Paleoclimatic and archeological implications" *Climatic Change* 9 (1986): 353.

[6]Minze Stuiver and Bernd Becker, "High-Precision Decadal Calibration of the Radiocarbon Time Scale, AD 1950–6000 BC," *Radiocarbon* 35.1 (1993): 55.

Conclusion

Salt-covered land snail shells at Qumran do not falsify the claim that Noah's Flood extended to Palestine, filling the Dead Sea depression 3520±21 B.C.

Chapter Eleven

Noah's Flood in Ireland

From a global perspective, Palestine and South Mesopotamia are relatively close to one another. Let us open the field of view to a much larger segment of the planet. Let us inquire after Noah's Flood within the secular archaeological data of Ireland (Figure 11.1).

Figure 11.1: Location of Ireland relative to Palestine and South Mesopotamia.

The distance between Ireland and South Mesopotamia is roughly five times greater than the distance between Palestine and South Mesopotamia, spanning about one eighth (12.5%) of earth's circumference. Is there any evidence of Noah's Flood within the secular scientific record as far removed from South Mesopotamia as Ireland?

Indeed there is.

Chronology of Ireland

In Mesopotamia and Palestine, the task of constructing a chronology of the distant past was greatly assisted by the observed stratigraphy at numerous archaeological sites. The site of ancient Jericho in Palestine, for example, is like a layer cake, with differing strata piled on top of one another, the total spanning thousands of years of human occupation. The presence of strata not only revealed to the archaeologists, early on, the proper ordering of the various cultural assemblages normal to that region, but the thicknesses of the strata could even be used to make crude estimates of the duration of the corresponding cultures at a particular site.

Nothing like Jericho exists in Ireland. Mesopotamia and Palestine have been fairly densely populated for a very long time. Not so Ireland. Archaeological remains from the middle of the fourth millennium B.C. are relatively rare in Ireland. As a result, chronology-building has been an unusually difficult enterprise in the archaeology of Ireland.

I have several books in my personal library dealing with the history and prehistory of Ireland. None of those books contains the sort of chronological table normal to books on the history and prehistory of Mesopotamia and Palestine. Were it not for radiocarbon dating, construction of a reasonably accurate chronology of Ireland's past seven or eight thousand years would be extremely difficult if not altogether impossible. Fortunately, because of radiocarbon dating, reasonably accurate chronologies of ancient Ireland can be constructed with relative ease today.

Figure 11.2 shows one such chronology. It is not a chronology of archaeological cultures. Rather, it is a chronology of forest versus field. But it has much to say which is of interest to the study of the archaeology of Ireland in relation to Noah's Flood, as will be seen.

The Figure 11.2 chronology results from the labors of two scientists,

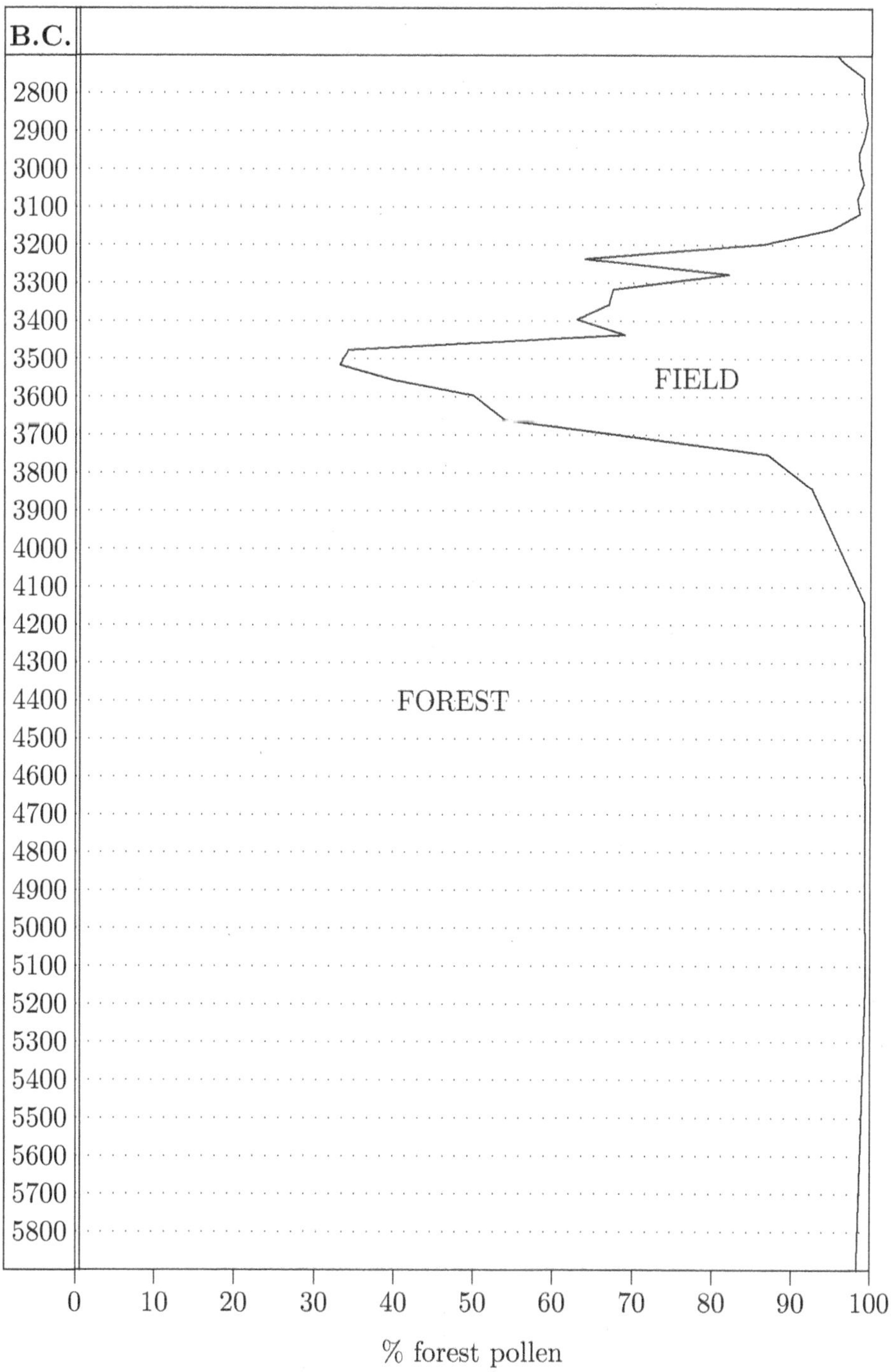

Figure 11.2: Composite pollen curve from a deep bog in Ireland indicating the relative abundance of forest and field in the vicinity of the bog.

Karen Molloy and Michael O'Connell, of the Palaeoenvironmental Research Unit, Department of Botany, University College Galway, Ireland.[1] Their science is known as palynology.

Palynology deals with pollen and spores. Pollen is interesting because it reveals the types of plants which were covering the countryside at different times. Each type of plant produces its own unique pollen.

The Figure 11.2 chronology results from pollen counts on peat samples cored from a deep peat bog in northwestern Ireland. For Figure 11.2 the total pollen count has been divided between forest types of pollen and field types of pollen. The "forest" types of pollen are mainly pine and other trees with a generally minor contribution of tall shrubs and ferns. The "field" types of pollen are from the herb family, including cereals and other grasses as well as typical field weeds such as ribwort, dock, buttercup, and dandelion.

The cost in money and man-hours to construct such a chronology is obviously not small. Construction of this particular chronology involved:

1. locating the deep bog,
2. making numerous trial corings at the site to determine where the deepest, most ancient part of the record might be found,
3. obtaining a ten centimeter diameter core some five and a half meters (roughly seventeen feet) long from top to bottom of the bog,
4. sectioning the core back at the lab to obtain more or less evenly spaced, one centimeter thick samples for analysis,
5. submitting each of these one centimeter samples to an elaborate process to separate the pollen it contained from everything else which might be found in a peat bog,
6. mounting each resultant pollen sample on a microscope slide,
7. identifying the individual type of up to 1000 pollen spores from each sample by viewing at 500× through the microscope,

[1] Karen Molloy and Michael O'Connell, "Palaeoecological investigations towards the reconstruction of environment and land-use changes during prehistory at Céide Fields, western Ireland," *Probleme der Küstenforschung im südlichen Nordseegebiet* 23 (1995): 187–225.

8. graphing the frequency of the different types of pollen versus the depth from which the sample was obtained in the peat core, and, finally,

9. submitting samples of peat from the core for radiocarbon dating so a conversion from depth of peat to calendar years could be worked out.

It is, of course, a great privilege to have such data available today. These sorts of chronologically controlled datasets have become available only in the past several decades. There was nothing comparable in earlier decades.

Based upon the number and precision of radiocarbon dates made on peat samples from this core, the conversion from depth of peat to calendar years may be estimated to be accurate to probably ±150 years back to roughly 4000 B.C., with larger uncertainties before that date.

Archaeology of Ireland

The Figure 11.2 chronology reveals the clearing of virgin forest by early settlers of Ireland. It shows that for millennia prior to about 4100 B.C. there was almost exclusively forest pollen, with very little pollen from typical field plants. But then the forest began to be driven back, and fields began to take its place.

The first settlers of Ireland, like the early European settlers of the east coast of North America, found a forested land when they first arrived there. They set about to clear the forests to make agricultural fields, just as the early European settlers of North America did.

The success of this settlement process can be followed by observing the increasing percentage of field pollens and declining percentage of forest pollens from 4100 B.C. onward. The Figure 11.2 chronology reveals that clearance of the forest increased at a more or less exponential rate for some six hundred years. But then, *in the middle of the fourth millennium B.C.*, there occurred a sudden reversal. Starting near 3500 B.C., the fields rapidly began to disappear, and the forest began to re-establish itself. Four hundred years later, by 3100 B.C., the fields had been fully overgrown by trees once again.

Here then, through the humble agent of microscopic pollen spores, a record of an ancient, human-instigated battle between forest and field in Ireland has been carefully archived for multiple millennia within the soggy peat of a deep bog. By exploiting this ancient archive, the arrival of the first settlers in Ireland can be observed, the increasing prosperity of their settlement can be followed for some six hundred years (to appreciate all that might be accomplished by an initial settlement after such a lapse of time, it may help to recall that Columbus discovered the Americas just over five hundred years ago), and a pretty serious reversal can be seen to begin roughly 3500 B.C.—synchronous with Noah's Flood.

Céide Fields

Pollen is hardly the only thing from the archaeology of Ireland which is of interest to the present investigation. The peat bog from which the Figure 11.2 chronology was derived is located in close proximity to a vast ancient field system. So special is this ancient field system that a multi-million dollar tourist center was built in the early 1990s to show it off. The fields are a major tourist attraction, with tens of thousands of visitors annually. Molloy and O'Connell chose to study the bog they did because of the light they knew it would shed on these ancient fields.

The field system is known today as "Céide Fields" (pronounced kay'jeh). The fields are delineated by miles of ancient stone walls. They cover at least 1000 hectares (or roughly 2500 acres). They constitute a rare, prehistoric landscape. This landscape has been preserved because it has been buried beneath a thick blanket of peat for thousands of years.

There can be no doubt about the original purpose of these fields. They were obviously used for agriculture. They go back to a time, labeled the Irish Neolithic by the archaeologists, when the farmers who worked the land lacked the advantage of metals. They made their tools entirely of chipped stone, bone, and wood. Axe heads fashioned from igneous rock have been found in association with these stone walls, and flint arrowheads (Figure 11.3), and flint scrapers, but never any metals.

The fields are not laid out willy-nilly. They give evidence of large-scale planning and organized effort. In these fields and in these walls, it is possible to begin to get to know something of the settlers whose activities are responsible for the pollen chronology of Figure 11.2—of the ancient farmers who cleared the forests and established the fields.

Many fascinating questions arise in connection with these farmers. Who were they? Where had they come from? How did they live?

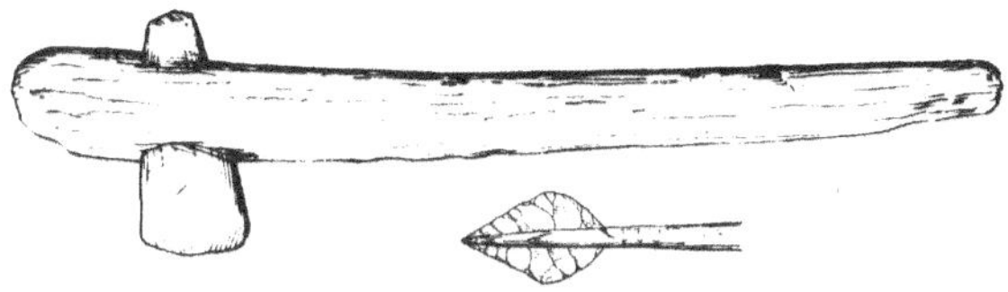

Figure 11.3: Typical tools of the Irish Neolithic: hafted stone axe head and arrowhead. [See Michael Herity and George Eogan, *Ireland in Prehistory* (London: Routledge & Kegan Paul, 1977), 41 for original credits.]

But there is one question in connection with them which overshadows all others. *What happened to them?*

Like the streets of Pompeii beneath their blanket of ash, the ancient agricultural landscape of Céide Fields lies preserved beneath a blanket of peat. But, unlike Pompeii—the fossilized forms of whose inhabitants can still be recovered from its blanket of ash today—the inhabitants of Céide Fields are entirely missing. Where did they disappear to? How did they come to leave their fields—their homeland—to the creeping depredations of desolate bog and silent forest?

Archaeologists have been probing these ancient fields, asking these same questions for some decades now. They talk about the "abandonment" of Céide Fields and speculate on what prompted it.

Some have suggested that the climate may have changed, forcing the occupants to leave. Maybe it just got too wet, encouraging rapid bog growth and making the land just too difficult to farm.

Others have supposed that loss of soil fertility was the culprit. They point out that the clearing of the original forests for agricultural purposes would have exposed the soil directly to the abundant rainfall in that region. Perhaps this led to a leaching away of soil nutrients.

Or maybe the farmers allowed their cattle to overgraze the land, robbing it of grass cover and inviting the formation of peat from mosses, and hence turning pasture to bog.

These are all interesting suggestions. But several facts argue against them. For example, there is the fact that farming resumed in the same region not later than about 2400 B.C. with a renewed clearance of forest—once again revealed by pollen from peat cores. This farming carried on for several thousand years without interruption. So there seems no reason to believe farming is ill-fated by nature in this region of Ireland.

And this later farming activity demonstrates unequivocally that there was clearly opportunity for the original farmers to carve new fields from adjacent forest, had loss of fertility from their centuries-old fields been the problem.

But most unsettling to these theories is the observed coincidence in time between the depopulation of these fields and the similar depopulation events already noted in Mesopotamia and Palestine. Check the archaeological record where you will, and you will find that such depopulation events are rare. The pollen record at Céide Fields, for example, never again shows a downturn in agricultural fields lasting a thousand years, or anywhere even close to a thousand years, as it did with this mid-fourth-millennium B.C. event.

Can three established cultures vanish at the same time and there be no underlying unifying cause? But surely it cannot be supposed that increasing rainfall, or loss of soil fertility, or overgrazing will function as an explanation of the observed depopulations simultaneously in all three of these geographically widely separated cases. These theories are found to be inadequate the moment one's field of view is enlarged from the local to the global scale.

Much simpler and more cogent, then, is the explanation that has passed down to us from this ancient time—that humanity was all but exterminated by a great Flood which covered high mountains and lasted for months.

Speed of "Abandonment" of Céide Fields

The Figure 11.2 pollen data showing shrinkage of the fields after 3500 B.C. allow two interpretations. It is possible to imagine a population which, having suffered a serious setback in roughly 3500 B.C., fights a losing battle with nature, slowly dwindling away to nothing over the subsequent four centuries. Alternatively, it is possible to imagine, in harmony with the written record of Noah's Flood found in Genesis, that the population was entirely wiped out in roughly 3500 B.C. and that the regression of the fields and regrowth of the forest took their natural course without further human intervention over the subsequent four centuries.

Karen Molloy and Michael O'Connell, the scientists responsible for the Figure 11.2 chronology, are probably in as good a position as anyone to speak to this question as a result of their careful studies at Céide Fields.

After noting the difficulty of answering such a question on the basis of paleoecological information, they nonetheless conclude "it appears that abandonment of the field system was rapid..."[2]

Conclusion

The inhabitants of Céide Fields, Ireland, vanished synchronously with the inhabitants of Palestine and with the inhabitants of South Mesopotamia in the middle of the fourth millennium B.C.

Synchronous depopulation over such a large geographical area is exceedingly rare in archaeology. Indeed, this may be its only instance.

Such a widespread depopulation event is not at all easy to explain. But Genesis provides an explanation. It describes a mountain-covering Flood which drowned almost everybody in the middle of the fourth millennium B.C.

[2]Karen Molloy and Michael O'Connell, "Palaeoecological investigations towards the reconstruction of environment and land-use changes during prehistory at Céide Fields, western Ireland," *Probleme der Küstenforschung im südlichen Nordseegebiet* 23 (1995): 219.

CHAPTER TWELVE

CHECK: PINE STUMPS

It is possible to carry out another quick check at this point. This time it is stumps of pine trees which furnish the check.

The stumps were found rooted in position of growth within the blanket peat covering the agricultural fields at Céide Fields and nearby areas. Because the stumps were rooted within the peat, rather than in the mineral soil of the underlying fields, it is clear that they only began to grow *after* the fields had gone out of use. This provides a simple means of checking the claim that Noah's Flood is responsible for the depopulation of Céide Fields.

Specifically, if the pine forest, of which these stumps are a residual, resulted from natural overgrowth of agricultural fields due to depopulation of Céide Fields by Noah's Flood, then these stumps should date to within a few centuries *following* the Flood. If they are found to date to any time *prior to* the Flood (i.e., prior to 3520±21 B.C.) they will entirely falsify the claim that Noah's Flood is responsible for the depopulation of Céide Fields.

The Data

Figure 12.1 shows radiocarbon dates from forty pine stumps found preserved within the blanket peat of Céide Fields and nearby areas.[1] Each

[1]Seamas Caulfield, R. G. O'Donnell, and P. I. Mitchell, "^{14}C Dating of a Neolithic Field System at Céide Fields, County Mayo, Ireland," *Radiocarbon* 40.2 (1998): 629–640. Calibration of these radiocarbon dates was carried out using the bidecadal dataset

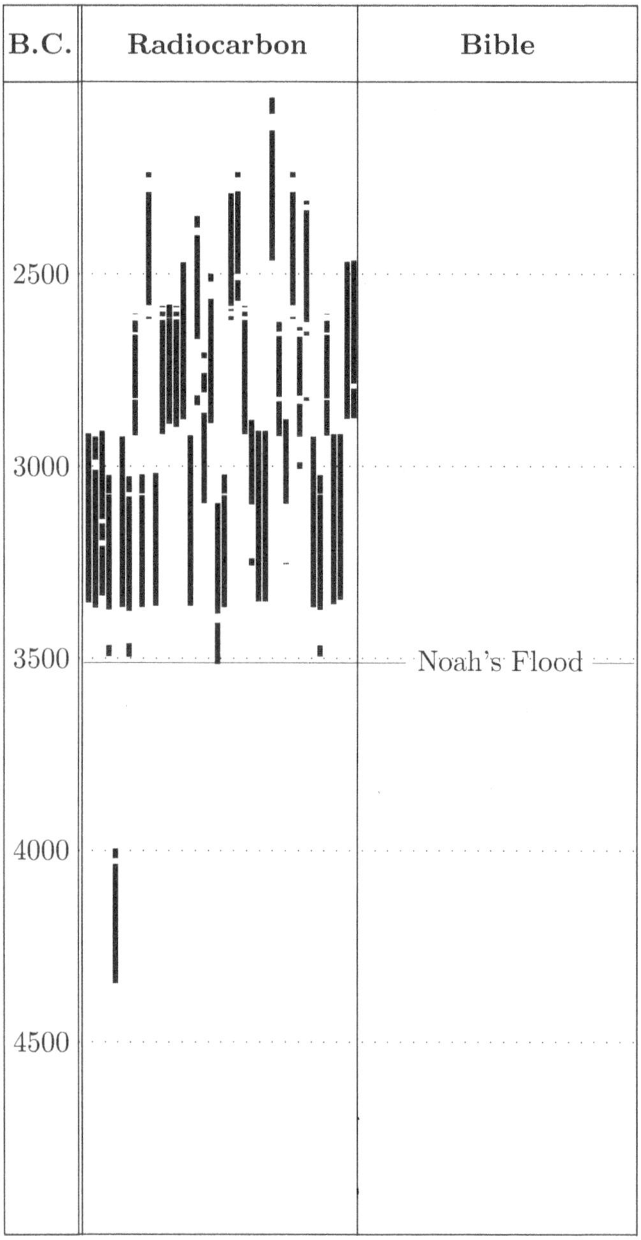

Figure 12.1: Radiocarbon dates on tree stumps from Céide Fields relative to the modern biblical chronology date of Noah's Flood.

(often broken) vertical black bar indicates the interval(s) in which the true calendar date of the sample is most likely to lie (i.e., when the wood sample being dated most likely grew). The probability is roughly two-thirds that the true date of the sample falls somewhere in the black barred region(s).

What is immediately striking about Figure 12.1 is that all but one of these stumps date after the Flood, and many date relatively soon after the Flood. (The one older, pre-Flood date—barring any sort of dating error—would reflect a tree which grew before clearance of Céide Fields had even commenced.) A cause-and-effect relationship between the Flood and the incursion of these trees is obviously strengthened by these data.

Figure 12.1 shows that trees were well established over Céide Fields by 3200 B.C. at the latest. This is consistent with the view that these fields ceased to be used because the region was depopulated by Noah's Flood 3520±21 B.C. With nobody there to tend the fields any longer, Céide Fields would have overgrown first with weeds, then with brush, and eventually with pine forest.

Conclusion

The data displayed in Figure 12.1—radiocarbon dates on ancient pine stumps from Céide Fields—had potential to falsify the claim that Céide Fields was depopulated by Noah's Flood, but they did not do so. Rather, they went in the opposite direction, corroborating and strengthening the claim by showing that by 3200 B.C., Céide Fields wore a veneer of blanket peat and pine trees over vast fields which had been cultivated for agricultural purposes three centuries previously.

On the basis of the radiocarbon dates found for these pine stumps, the original researchers, S. Caulfield *et al.*, concluded:[2]

> At the latest, blanket peat was widespread in North Mayo by 4500 BP [3265 B.C.]. Indeed, it is reasonable to suggest that

of CALIB 3.0.3. See M. Stuiver and P. J. Reimer, "Extended ^{14}C Data Base and Revised CALIB 3.0 ^{14}C Age Calibration Program," *Radiocarbon* 35.1 (1993): 215–230.

[2]Seamas Caulfield, R. G. O'Donnell, and P. I. Mitchell, "^{14}C Dating of a Neolithic Field System at Céide Fields, County Mayo, Ireland," *radiocarbon* 40.2 (1998): 629–640.

> it was widespread some 500 years earlier, given the depth of peat beneath some of the pine stumps...

Thus, these researchers conclude that the initiation of blanket peat over Céide Fields should date somewhere between the extremes of 3765 B.C. and 3265 B.C. This range can be written succinctly as 3515±250 B.C. This secular date for the initiation of blanket peat over Céide Fields is indistinguishable from the modern biblical chronology date of Noah's Flood, 3520±21 B.C.

Thus the pine stumps corroborate the findings of the previous chapter: the inhabitants of Céide Fields were swept away by Noah's Flood.

CHAPTER THIRTEEN

OBJECTION: THE BIBLE TEACHES THAT THE FLOOD WAS A CATACLYSM

Evidence of Noah's Flood has been looked for and found in South Mesopotamia. Evidence of Noah's Flood has been looked for and found in Palestine. And evidence of Noah's Flood has been looked for and found in Ireland.

These are not the end of presently available evidences for Noah's Flood in the middle of the fourth millennium B.C., by any means, as subsequent chapters will show. But it is necessary to pause here and reflect briefly on what just these three evidences are saying about the nature of the Flood.

They are, of course, seriously damaging to the idea that Noah's Flood is but a myth. The simultaneous vanishing of three geographically separated civilizations, at the very time called for by the Genesis Flood narrative, is difficult to reconcile with the myth hypothesis. But I will not press this point here. The historical factuality of Noah's Flood will be proven beyond reasonable doubt in subsequent chapters. More pressing at present is a concern of those who already accept the Genesis Flood narrative as historical, but who have been taught to believe a fanciful interpretation of that narrative.

The three instances of the Flood which have been looked at so far show clearly that Noah's Flood was not a local flood, nor even a regional flood.

It extinguished civilizations an eighth of earth's circumference apart. It reached beyond continental boundaries, crossing the Mediterranean Sea and a portion of the Atlantic Ocean. It was evidently a massive affair, a catastrophe of a type unsuspected by modern earth science.

But it was not an earth-shattering tectonic cataclysm. It was not responsible for most or even any significant portion of the geological column or its entombed fossils. It did not rip up and redeposit the surface of the earth. The intact archaeological strata in Mesopotamia and Palestine demonstrate this, and the pre-Flood stone walls at Céide Fields absolutely guarantee it.

And this raises a final objection which must be dealt with before concluding Part I of this volume. This is the objection that the Bible teaches that Noah's Flood was a tectonic cataclysm of such ferocity that no trace of pre-Flood civilization could possibly have survived it.

General Background

In 1961 theologian John C. Whitcomb, Jr. and engineer Henry M. Morris claimed Noah's Flood "was a gigantic catastrophe, beside which the explosion of the largest hydrogen bomb, or of hundreds of such bombs, becomes insignificant!"[1] They argued that the biblical text and the book of nature clearly portrayed this historic event as a cataclysm—a great overwhelming geologic upheaval. They pictured the Flood as accompanied by great tectonic events, earthquakes, volcanic eruptions, and tidal waves which together wreaked unimaginable havoc upon the face of the globe world-wide.[2] They claimed it produced most of the layers of sedimentary rock strata which are found around the world, some of which are over a mile deep.[3]

Whitcomb and Morris felt the Bible guaranteed their view of Noah's Flood. This led them to sometimes insinuate heresy against any who disagreed with them. They claimed, for example:[4]

[1] John C. Whitcomb, Jr. and Henry M. Morris, *The Genesis Flood* (Philadelphia: The Presbyterian and Reformed Publishing Company, 1961), 242–243.

[2] John C. Whitcomb, Jr. and Henry M. Morris, *The Genesis Flood* (Philadelphia: The Presbyterian and Reformed Publishing Company, 1961), 122–123, 261, 264–265, 268–269, 271.

[3] John C. Whitcomb, Jr. and Henry M. Morris, *The Genesis Flood* (Philadelphia: The Presbyterian and Reformed Publishing Company, 1961), 123, 265–266, 268–272.

[4] John C. Whitcomb, Jr. and Henry M. Morris, *The Genesis Flood* (Philadelphia:

> There is no escaping the conclusion that, if the Bible is true and if the Lord Jesus Christ possessed divine omniscience, the Deluge [Noah's Flood] was the most significant event, geologically speaking, that has ever occurred on the earth since its creation.

That is a pretty strong claim. Is it true? Does the Bible teach that the Flood was a cataclysm—an event characterized by violent tectonic upheaval and demolition of the surface of the earth?

Hermeneutical Considerations

While it is easy to see why it might be argued that the Bible teaches the Flood was *global*, it is much less obvious why it should be accepted that the Bible teaches the Flood was *cataclysmic*. It is unquestionably the case, in apparent contradiction to Whitcomb and Morris' claim above, that Jesus is nowhere recorded in the Bible as having said the Flood "was the most significant event, geologically speaking, that has ever occurred on the earth since its creation," and He makes no mention of earthquakes, tidal waves, volcanoes, or mile-deep sediments at the time of the Flood that I have been able to find.

The biblical historical record of Noah's Flood found in Genesis chapters 6 through 9, where such teaching might naturally be looked for, also contains no explicit reference to earthquakes, volcanoes, tidal waves, or mile-deep sediments. This is a strange silence if this event was, in fact, not just accompanied by, but, indeed, *characterized* by such phenomena, as Whitcomb and Morris claim. The biblical record is, after all, quite detailed, and even quantitative, regarding the *water* of the Flood. It states where the water came from, how long the rain lasted, how it lifted the ark, how deep the water became, how it covered everything in sight, how long it continued to rise, how it receded, and how long it took to dry up. Why so much detail about the water but complete silence regarding the claimed tidal waves, earthquakes, and volcanoes?

Genesis does say that at the beginning of the Flood "all the fountains of the great deep burst open"[5], and Whitcomb and Morris make a great deal of this phrase. They say:[6]

The Presbyterian and Reformed Publishing Company, 1961), 216.

[5]Genesis 7:11

[6]John C. Whitcomb, Jr. and Henry M. Morris, *The Genesis Flood* (Philadelphia:

> This must mean that great quantities of liquids, perhaps liquid rocks or magmas, as well as water (probably steam), had been confined under great pressure below the surface rock structure of the earth since the time of its formation and that this mass now burst forth through great fountains, probably both on the lands and under the seas.

Why this simple phrase "must" mean this is not at all clear to me. Certainly the context of the phrase (i.e., Genesis 6–9) provides no hint of subterranean reservoirs of molten rock and steam erupting at this time. The context restricts to just plain, ordinary water flooding the earth.

So I find Whitcomb and Morris' exegesis of the phrase, "all the fountains of the great deep burst open," difficult to accept. In fact, their hermeneutical approach to the entire verse seems unsound to me. Notice that they interpret "fountains of the great deep" literally. They say that these mean "great fountains, probably both on the lands and under the seas." But then they go on to interpret the parallel phrase, "and the floodgates of the sky were opened," which immediately follows in the same verse, metaphorically. They say:[7]

> Speaking metaphorically, the Scriptures say that the "floodgates of heaven were opened."

Surely it is not sound hermeneutics to interpret the first of a set of parallel phrases literally, and the second metaphorically, is it? If the "fountains of the great deep" correspond to literal, physical fountains, shouldn't "the floodgates of the sky" correspond to literal gates blocking water behind literal, physical dams in the sky? And if the notion of literal "floodgates of the sky" seems sufficiently absurd as to demand a metaphorical interpretation, shouldn't that immediately tell us that the parallel "fountains of the great deep" should also be understood metaphorically? Wouldn't it be far more reasonable to see in these two phrases simply the idea that the water which caused the Flood came as rain from the sky and as a transgression from the sea?

In any event, this little phrase certainly seems inadequate justification of Whitcomb and Morris' claim that "if the Bible is true and if the

---

The Presbyterian and Reformed Publishing Company, 1961), 122.

[7]John C. Whitcomb, Jr. and Henry M. Morris, *The Genesis Flood* (Philadelphia: The Presbyterian and Reformed Publishing Company, 1961), 120.

Lord Jesus Christ possessed divine omniscience, the Deluge was the most significant event, geologically speaking, that has ever occurred on the earth since its creation." Yet I can find precious little else in their book, *The Genesis Flood*, to justify this claim.

Indeed, when Whitcomb and Morris finally set about to develop their cataclysmic Flood model for their reader, they do so, not on the solid basis of unambiguous, explicit biblical statements, but merely on the basis of *inference*. Under the heading "GEOLOGICAL IMPLICATIONS OF THE BIBLICAL RECORD" they write:[8]

> The only proper place to begin this study is with the Bible record of the Flood itself. The following appear to be legitimate *inferences* from the account: [my emphasis]

I am a scientist by training, not a theologian, so I have no intention of dissecting the fine points of what might or might not be *inferred* from the biblical text. I only wish to establish the point that the Bible nowhere explicitly teaches that the Flood was a cataclysm. Quite plainly, the *only* way Whitcomb and Morris have been able to arrive at their cataclysmic Flood model is through *inference*.

Now the fact that a cataclysmic Flood model *can* be inferred from the Bible does not mean that it *must* or *should* be inferred from the Bible. It is possible, through mistaken reasoning, to infer all sorts of things from the Bible which it simply does not teach. Please note that other investigators have arrived at quite different models for the Flood from their reading of the Bible. Evidently, one can honestly *infer* a variety of Flood models from the Bible—from cataclysmic to tranquil.

Psalm 104:5–9

But what about Psalm 104:5–9? Don't these Bible verses teach that the Flood was accompanied by the uplifting of mountains? And doesn't this explicitly show that the Flood was a great tectonic event?

Whitcomb and Morris write, for example:[9]

---

[8]John C. Whitcomb, Jr. and Henry M. Morris, *The Genesis Flood* (Philadelphia: The Presbyterian and Reformed Publishing Company, 1961), 120

[9]John C. Whitcomb, Jr. and Henry M. Morris, *The Genesis Flood* (Philadelphia: The Presbyterian and Reformed Publishing Company, 1961), 121–122.

> Very likely, in order to accommodate the great mass of waters and permit the land to appear again, great tectonic movements and isostatic adjustments would have to take place, forming the deep ocean basins and troughs and elevating the continents. This seems to be specifically implied in the poetic reflection of the Deluge in Psalm 104:5–9.

The passage in question reads (NASB 1975):

> 5. He established the earth upon its foundations,
> So that it will not totter forever and ever.
> 6. Thou didst cover it with the deep as with a garment;
> The waters were standing above the mountains.
> 7. At Thy rebuke they fled;
> At the sound of Thy thunder they hurried away.
> 8. The mountains rose; the valleys sank down
> To the place which Thou didst establish for them.
> 9. Thou didst set a boundary that they may not pass over;
> That they may not return to cover the earth.

This passage does seem to be referring to the Flood, but its interpretation in terms of tectonics seems to me to be questionable for a number of reasons.

Notice, first of all, that it is not this whole section of five verses which can be appealed to for support of rising continents and sinking ocean basins (i.e., tectonics) but just one half of one verse. Specifically, what is utilized is the first half of verse 8, which says, "The mountains rose; the valleys sank down." But even this half does not say the *continents* rose and the *ocean basins* sank down. It says the *mountains* rose and the *valleys* sank down. Obviously, it is a rather large leap to change what the text calls "mountains" into continents, and "valleys" into ocean basins. But I recognize that care must be exercised with how literally a phrase from poetic verse should be taken, so I will not press this point.

Of greater concern is the fact that the context doesn't seem to support a tectonic interpretation of this passage. The subject of verses 6, 7, and 9 is the water of the Flood, not the mountains or the valleys. The Bible reader is watching the Flood waters in these verses. He is watching the waters in action as they respond to the sovereign voice and will of God. In verse 6 the waters are covering the earth. In verse 7 they are fleeing. In verse 9 they are restrained. It seems foreign to interject moving mountains and valleys in the middle of this focus on the waters, and I believe the poet did not intend that we should.

I suggest that the psalmist is not talking about any absolute motion of the mountains and valleys at all here. I suggest that he is simply carrying on with his theme of the waters in motion in response to the will of God. What the poet is describing in verse 8 is not what the mountains and valleys are doing. Rather, he is describing what is seen as a result of what the waters are doing.

The waters are retreating. As a result, looking out from the ark for example, the mountains appear to be emerging from the water more and more each day. Similarly, as the surface of the water sinks lower and lower each day, the valleys between the mountains appear to be deepening.

I suggest that it is this concept which underlies verse 8, rather than any absolute motion of mountains and valleys. The psalmist has already made it clear in verses 6 and 7 that it is the waters which are doing the moving. That the waters have not been set aside as the subject in verse 8 is clear by the fact that they are still the subject of verse 9. Thus the action of the waters should still be regarded as the subject of verse 8, not the action of the mountains and the valleys.

Psalm 104:5–9, which is poetry after all, cannot legitimately be used to say that the Bible teaches the Flood was a great tectonic affair.

Conclusion

The cataclysmic Flood model is *not* a Bible doctrine by any legitimate exposition of Scripture. It is a *scientific* model only, on the same plane as every other model of Noah's Flood which has ever been inferred from the Bible. It can legitimately claim no special supra-scientific status. And since this is the case, it is altogether proper and appropriate to submit this model to the usual rigors of scientific examination and to declare it false when it is found to fail the test—with no consequent aspersions on the divinity of Christ or the truth of the Bible.

Chapter Fourteen

Conclusion to Part 1

The biblical record makes extraordinary claims about the Flood and its aftermath. These claims require near-extermination of humanity over a large portion of the globe. Nothing remotely matching this requirement is found at the traditional biblical chronology date of the Flood, in the middle of the third millennium B.C. But in the middle of the fourth millennium B.C., where modern biblical chronology says the Flood should be found, evidence fulfilling this requirement is immediately available from secular disciplines.

It doesn't take a genius to understand what is going on here. It is really very simple. A copy error was made on a single number in a single verse of the Bible (1 Kings 6:1) a very long time ago. This small copy error had an unfortunately large impact on biblical chronology. It resulted in the loss of exactly one thousand years from biblical chronology prior to the time of Samuel. As a result, archaeologists and other scholars mistakenly supposed they should find the Flood in the middle of the third millennium B.C. When, after many decades, efforts to find the Flood in the middle of the third millennium B.C. had uniformly failed, most scholars concluded that the Flood was not real history—that it was a myth.

But that conclusion is utterly wrong. The biblical narrative of the Flood accurately portrays a real historical event. *All* that is wrong is traditional biblical chronology. Restore the accidentally dropped "and thousand" to its rightful place in 1 Kings 6:1 (page 36, Figure 3.1), thus restoring the proper date of the Flood to the middle of the fourth millennium B.C., and the Flood is immediately apparent in the secular archaeological data, as has begun to be shown.

Surely the time has come to reexamine scientific attitudes toward the biblical Flood narrative. The archaeological evidence is clear that something extraordinary took place on Earth five and a half thousand years ago. Widely separated civilizations vanished without a trace. The ancients have obviously preserved a memory of this event through several written sources, including that which is found in Genesis. They have clearly attempted to communicate that what took place was a flood of nearly unimaginable proportions, in size, duration, and consequence.

But our generation, having come of age in the midst of an explosion of scientific knowledge, has been too "smart" to listen. Our generation has so far chosen to "correct," ignore, or even to ridicule these ancient memories.

Is it possible that our reaction may be revealing a certain "scientific adolescence" on our part? Is it possible that it has been our immature scientific knowledge which has kept us from being able to understand and appreciate what our predecessors have been trying to tell us all along? Is it possibly time we *moderns* grew up a bit?

Part 2

The Genesis Record of Noah's Flood

CHAPTER FIFTEEN

THE DATE AND ORIGIN OF THE GENESIS FLOOD ACCOUNT

It has often been claimed that the Genesis account of the Flood is a late derivative of the Gilgamesh Epic. Neev and Emery make this claim, for example, in their previously mentioned book, *The Destruction of Sodom, Gomorrah, and Jericho*:[1]

> The latter dates are near enough to the time of the epic flood described in *Gilgamesh* ... The Babylonian Utnapishtim, Gilgamesh's ancestor, was instructed by Ea, Sumerian Enki or Lord of the Earth, to build a ship and load it with the seed of all living things, cattle, beasts, and his family. According to the story, the ship survived the flood about 6000 B.P. [*ca.* 4000 B.C.] and landed on Mount Nisir. This account must have been current at Uruk, the religious capital where Gilgamesh was king and known to Abraham who lived at Ur—now Eridu—only 60 km distant until he emigrated to Canaan. Evidently the story of Utnapishtim's flood evolved into the later and more detailed story of Noah's Flood.

Woolley also makes this claim in his book, *Spadework in Archaeology*:[2]

[1]David Neev and K. O. Emery, *The Destruction of Sodom, Gomorrah, and Jericho*, (New York: Oxford University Press, 1995), 119.

[2]Sir Leonard Woolley, *Spadework in Archaeology* (New York: Philosophical Library, 1953), 105-106.

> Under the eleven-foot stratum lay the ruins of the houses in which had lived the antediluvian inhabitants of the Lower Town; they had, we may suppose, taken refuge on the mound, the Inner City, and from its walls watched their homes disappear beneath the muddy waters of the flood. This was the flood, coming in the latter part of the al 'Ubaid period, which the ancient Sumerians regarded as the outstanding disaster in their country's history, and out of the historic fact grew the legend which in course of time the Hebrew people incorporated in their own sacred writings and handed on to us today, the story of Noah's Flood.

Traditional biblical chronology lends considerable assistance to this claim. Modern biblical chronology says it is false.

Traditional biblical chronology placed the Flood near 2500 B.C., and Abraham in the 21st century B.C. These dates seem quite late relative to Gilgamesh, who is presently believed to have ruled in the 27th century B.C.

But modern biblical chronology, which restores the missing thousand years to 1 Kings 6:1 (page 36, Figure 3.1), reverses all this. It places the Flood 3520 B.C., and Abraham in the 31st century B.C., long before the reign of Gilgamesh.

With this temporal arrangement, the thesis that Genesis preserves a historical record of the Flood, later distorted for literary purposes in the Gilgamesh Epic, must be taken seriously.

It doesn't take a degree in ancient literature to recognize that the Genesis narrative of the Flood does not have the feel of legend about it. Rather, with its carefully recorded calendar of events and observations throughout the year of the Flood (next chapter), it feels like a personal journal—a diary, a ship's log, and a science lab notebook (e.g., the record of the bird experiments following the Flood[3]) all rolled into one. It contrasts starkly with the Gilgamesh Epic, whose blatant carnality and fast-paced, vainglorious exploits clearly signal phantasm.

From my years of study of the Genesis Flood account *as a physicist*, it seems about as certain as a thing can be that the Genesis Flood narrative cannot be anything other than an eyewitness record of the phenomenal events it describes. The fundamental reason for this is that the Flood

[3]Genesis 8:6–12.

Figure 15.1: Cuneiform tablet. [See J.N. Postgate, *Early Mesopotamia: Society and Economy at the Dawn of History* (New York: Routledge, 1994), 57 for details and credits.]

event, as it is given in Genesis, is explicable on strictly scientific grounds, as I will show in subsequent chapters, but the basic science which renders it explicable has only come to be known in the past century.

I suggest that the account of the Flood which is preserved in Genesis is very likely primarily a transcription of Noah's written record of his own actions and observations, and that it should, therefore, be dated to the year of the Flood itself, that is to 3520±21 B.C. This is very near to the origin of writing in Mesopotamia, and it seems likely that the narrative was originally written in cuneiform on clay tablets (Figure 15.1). I suggest, in fact, that the biblical Flood narrative is the oldest account which we possess on Earth of a real historical event to have been recorded in writing at the time it happened.[4]

[4]Yet older events in Genesis give every impression of having been preserved orally prior to the invention of writing. See Gerald E. Aardsma, "The Origin and Antiquity of the Biblical Text," *The Biblical Chronologist* 8.6 (November/December 2002): 1–9. www.BiblicalChronologist.org.

CHAPTER SIXTEEN

CHRONOLOGY OF NOAH'S FLOOD

The narrative of the Flood found in Genesis chapters 7 and 8 contains a number of time references. For example, Genesis 7:11 (NASB 1975, my emphasis):

> **In the six hundredth year of Noah's life, in the second month, on the seventeenth day of the month,** on the same day all the fountains of the great deep burst open, and the floodgates of the sky were opened.

Such time references are normal to historical narrative and are not normally a part of myth or legend. The Gilgamesh Epic, for example, is devoid of anything similar.

Noah's Calendar

These time references constitute the basis for a historical chronology of the Flood event. However, these are obviously not references to the Gregorian calendar, which we now employ for keeping track of days, weeks, months, and years, since it was only in A.D. 1582 that the Gregorian calendar came into existence.

There have been many different calendars in antiquity. What calendar Noah used is not known. Most importantly for the purpose of the present chapter, it is not known how many days were in each of its months or how many months comprised a year.

These basic unknowns give rise to an unavoidable uncertainty in any attempt to fix the events of the Flood on a timeline. However, variations

Figure 16.1: The relative positions of sun, moon, and earth during the new moon phase. The rotational axis of the earth is perpendicular to the page, and the figure is viewed from above the North Pole. Not to scale.

in ancient calendar systems of the Middle East are sufficiently small to suggest that even in the worst case a cumulative error of only one or two weeks should be expected from any reasonable modern rendering of the chronology of the Flood.

While the lengths of a year or of the months in Noah's calendar are not known, it seems highly probable that they were linked to observed natural phenomena. "Day" is naturally defined as a single period of light and darkness caused by the rotation of the earth on its axis. "Month" finds its most natural definition in the revolution of the moon about the earth, a new month beginning with the first appearance of a crescent moon at dusk following its complete absence at night due to its close alignment with the sun (Figures 16.1 and 16.2). And similarly, "year" finds its most natural definition in the revolution of the earth about the sun, which gives rise to the observed annual cycle of seasons.

These natural definitions—which I will assume were the ones Noah used as I draft the following chronology of the Flood—though lacking in quantitative precision relative to modern technical scientific standards, would almost certainly function adequately for the day-to-day activities of people in the pre-Flood world. In fact, they seem to have been the original starting point from which all calendars of the Middle East in the post-Flood era were later devised. They are also obviously harmonious with the divine purpose, articulated at Creation, that the sun, moon, and stars should be "for seasons, and for days and years."[1]

These natural definitions have several interesting consequences. For

[1]Genesis 1:14. Notice that the moon plays only a minimal time-keeping role in Western calendrical practices today (associated with the timing of religious holidays), in contrast to the prominent role it played in many ancient calendar systems.

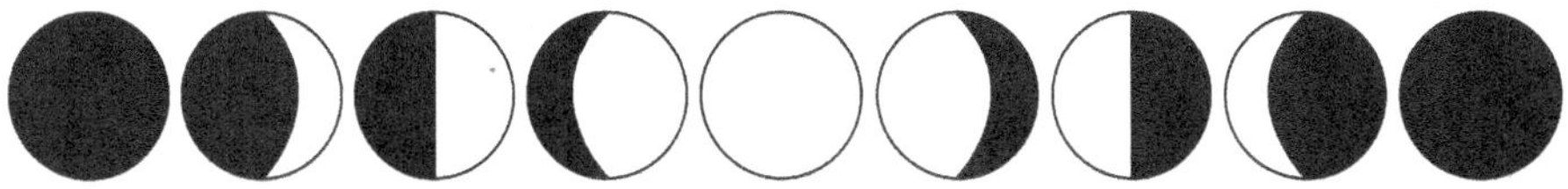

Figure 16.2: The phases of the moon from one new moon to the next. From left to right: new, crescent, first quarter, gibbous, full, gibbous, last quarter, crescent, and new. The first appearance of the crescent following the new moon is a point in this cycle which can be determined with relatively little ambiguity. Hence it serves as a good point from which to begin successive lunar months. At this point the moon is viewed at dusk, setting in the west, shortly after the sun has gone down.

example, they imply that the concept of "year" would be attached to the natural cycle of seasons rather than to any arbitrary count of days. Thus, the number of days comprising a year could vary slightly from year to year, but the long-term average would equal the mean solar year, which measures 365.2422 days at present. An important result is that the measured "years" of Noah's life recorded in the Bible, as well as those of the other pre-Flood patriarchs, would probably approximate solar years just about as closely as the "years" of our lives do according to current calendrical practices.

Unlike our modern calendar, however, one year would not divide neatly into twelve months. The observed period from one new moon to the next is called the synodic month. The synodic month is variable but averages 29.530588 days. Twelve synodic months equal 354.3671 days, which is 10.8751 days short of a solar year. Thus, in this natural system of reckoning time, months would not be expected to be synchronized with the solar year at all.

Such a calendar system, using lunar months as its basis, is called a lunar calendar. It is different from our modern solar calendar, which uses the solar year as its basis.

In a lunar calendar, with the beginning of each new month determined by observation, months vary in length between 29 and 30 days in an irregular way. However, the long-term average equals the average synodic month of 29.530588 days.

This all seems somewhat complicated, no doubt, but it amounts to a very simple and natural way of reckoning time in actual practice. For example, in this lunar calendar system, Genesis 7:11, quoted above, simply means that the Flood began on the seventeenth day following the second new moon to be observed by Noah (or his contemporaries) in his six hundredth year.

Throughout the remainder of this chapter, I will present such dates in the format "year of Noah's life/lunar month/day of month." For example, the calendar date in Genesis 7:11 is written in this format as 600/02/17.

Day Counts

Several "day counts" are used in place of calendar dates in the biblical narrative of the Flood: the initial forty days and nights of rain, the 150 days during which the Flood prevailed, and several counts in connection with the sending out of the birds from the ark. This is to be expected in such a natural calendar system because visibility issues would prevent observation of the new moon on some nights. Thus, the presence of such counts supports the suggestion that it was a lunar calendar which Noah actually employed.

Notice, for example, that it would not have been possible for Noah to observe the new moon during the forty days and nights of rain at the beginning of the Flood. Thus, while Noah would easily have been able to keep a running tally of how many days it had rained, he would not have been able to give an accurate lunar calendar date for when the rain stopped. The rain itself would have prevented observation of the new moon and, hence, it would have prevented determination of the exact day when the third month of Noah's six hundredth year began. I suggest that this may be the reason Noah recorded the number of days it rained, rather than the lunar calendar date when the rain stopped.

Visibility

Further support of the lunar, observational calendar thesis results from another, somewhat opposite observation. There are two calendar dates in the Flood narrative which, in a lunar calendar, would necessarily coincide with the appearance of the new moon: 600/10/01 in Genesis 8:5 and 601/01/01 in Genesis 8:13. If the lunar calendar thesis is correct, then atmospheric conditions on these two days would need to have been sufficiently clear for the new moon to be observed, to tell that the new month had begun.

In actual fact, in both instances observations are recorded which seem to imply clear conditions. Visibility was obviously good on 600/10/01, for Genesis 8:5 records that on that day the tops of neighboring mountains were seen. Similarly, on 601/01/01 Noah removed the covering of the ark and observed that "the surface of the ground was dried up." Presumably this "surface of the ground" is a reference to the plane below the mountain upon which the ark had come to rest, since Noah would already have known the surface of the ground on the mountain itself was dry, both from the appearance of the neighboring mountains whose tops had become visible two months previously, and from his dove experiments. Thus, good visibility is again implied, further corroborating the lunar calendar thesis.

Duration of the Flood

No matter which calendar Noah is assumed to have used, the total duration of the Flood was obviously about one year. The Flood began on the seventeenth day of the second month of Noah's six hundredth year (Genesis 7:11), and Noah and his family disembarked after the Flood on the twenty-seventh day of the second month of his six hundred and first year (Genesis 8:13–19).

Interestingly, however, when the natural calendar I have described above is assumed, the duration of the Flood comes out to be *exactly* 365 days. This, of course, is the length of the solar year (i.e., 365.2422 days) rounded to the nearest whole number of days.[2]

Thus it appears possible that the Genesis account of the Flood means to convey that Noah and his family were aboard the ark *exactly* one year.

This conclusion may find additional support from the Septuagint version of the Old Testament. Its reading of Genesis 7:11 and 8:14 is es-

[2]To see this, multiply the average synodic month of 29.530588 days by twelve to count the days from the beginning of the seventeenth day of the second month of Noah's six hundredth year to the beginning of the seventeenth day of the second month of Noah's six hundred and first year, then add eleven days to go from that point to the end of the twenty-seventh day. The result is $29.530588 \times 12 + 11 = 365.367056$ days.

I was introduced to this interesting fact several decades ago by a little tract called *Discovering the Calendar of the Creation*. [William G. Lowe, *Discovering the Calendar of the Creation* (Narrowsburg, New York: Scripture Truth, 1971).]

sentially the same as the reading of the Hebrew text (from which our English Bibles derive) with the single exception that in the Septuagint the Flood begins, not on the seventeenth day of the second month, but on the twenty-seventh day. This causes the Flood to begin and end on the same day of the same month of consecutive years. For those, like ourselves, who use a solar based calendar, events which begin and end on the same day of the same month separated by one year are immediately recognized to be one year long. I suggest the Septuagint reading in Genesis 7:11 may have been deliberately changed from "seventeenth day" of the Hebrew text to "twenty-seventh day" of the Septuagint for precisely this reason. I suggest that the Septuagint audience may have been familiar with a solar rather than a lunar calendar, and that Genesis 7:11 may have been changed in the Septuagint to communicate to this audience the fact that Noah and his family were aboard the ark exactly one year.

Modern biblical chronology finds the date Noah and his family disembarked from the ark to be 3519±21 B.C. (page 40). Taking the Flood to have begun exactly one year previously yields 3520±21 B.C. as the date of the beginning of the Flood.[3] These dates provide a functional chronological alignment of the Flood with respect to world history, as well as adequate boundaries for containment of the more detailed events of the Flood.

Detailed Chronology of the Flood

The events of the Flood recorded in Genesis chapters 7 and 8 are displayed in the time chart shown in Figure 16.3. The absolute (B.C.) dates discussed above are shown above and below the chart. This placement has been chosen to convey the idea that precise alignment of the chart with respect to the Gregorian calendar months is presently unknown.

The time scale for the chart is given in the leftmost column as a day number count. Noah's date references have been placed in the neighboring column to the right, at their appropriate day number. These are

[3]The quoted uncertainties on these two dates are understood to be correlated rather than independent. The fundamental biblical datum is that the Flood lasted one year, so the dates of commencement and termination of the Flood must always differ by one year only.

3519±21 B.C.

Day	Noah's Date	Event	Genesis
365	601/02/27	disembarking	8:14
		waiting	
310	601/01/01	covering of ark removed; water gone	8:13
283		dove sent out third time	8:12
276		dove sent out second time	8:10,11
269		dove sent out first time	8:8–10
262		raven sent out	8:6,7
		water still receding	
222	600/10/01	tops of mountains become visible	8:5
		water decreasing steadily	8:5
150	600/07/17	ark rests upon the mountains of Ararat	8:4
		water "prevailed"	7:24; 8:3
40		40 days and nights of rain ends	7:4,12,17
		raining; flooding	7:10–12,17–20
1	600/02/17	Flood begins	7:11

3520±21 B.C.

Figure 16.3: A chronology of Noah's observations of the Flood.

given in the format "year of Noah's life/lunar month/day of month" as mentioned above.

The Flood Begins

Genesis 7 opens with God's command to Noah and his family to enter the ark and board the animals. This command was given seven days prior to the beginning of the Flood.

Once Noah and his family had been shut safely inside, the Flood began on 600/02/17, evidently with nonstop rain ("the floodgates of the sky were opened"[4]) and the simultaneous rising up of water from the oceans onto the land ("all the fountains of the great deep burst open"[5]). This release of water over what had previously been dry land continued for forty days.

The Water Prevails

The word "prevailed" occurs three times in the last seven verses of Genesis chapter 7. This word is used each time in reference to the water of the Flood—we are told that the water prevailed 150 days. What does this mean?

The same Hebrew word is used of the battle between the Israelites and the Amalekites in the desert of Sinai after Israel had left Egypt. This battle, recorded in Exodus 17, is the one in which "it came about when Moses held his hand up, that Israel prevailed, and when he let his hand down, Amalek prevailed."[6]

I suggest that the use of "prevailed" in Genesis 7 has the same connotation as its use in Exodus 17. That is, there is a metaphor underlying the description of the Flood given in Genesis 7. The metaphor is of a conflict, not between warring armies, but between the land and the water. From the first day until the one hundred fiftieth day of the Flood the water appeared to be winning the battle, but from day 150 onward the water was in retreat.

The event which marked the turning point in this struggle appears to have been the onset of a drying wind (Genesis 8:1–3). The ark, by this time, was over the mountainous region where it would shortly come to

[4]Genesis 7:11; NASB 1975.
[5]Genesis 7:11; NASB 1975.
[6]Exodus 17:11; NASB 1975.

rest. Noah would have been unable to determine whether the water was rising or falling by taking soundings over such uneven terrain. But Noah would have been able to observe the onset of a dry wind, and if this wind persisted throughout the waning of the Flood, then it could have served as an observable indicator that the Flood had begun its retreat.

In addition, simple calendrical considerations require that the grounding of the ark on the mountain (Genesis 8:4) must have occurred on or very soon after day 150. The Genesis narrative records this grounding as the next event following the onset of the wind. It is recorded as having happened on 600/07/17. I have synchronized this calendar date with day 150 even though the text does not explicitly do so. This requires only that four of the five synodic months which completed between the start of the Flood on 600/02/17 and the grounding of the ark on 600/07/17 had thirty days and one had twenty-nine days. In his *Handbook of Biblical Chronology*, Jack Finegan displays a table of the actual month lengths recorded during the first nineteen years of the reign of Nebuchadnezzar II of Babylon.[7] I found twenty-four occurrences of five consecutive months in which four were thirty days long and the remaining one was twenty-nine days long in this table.

Once the ark had grounded Noah would have been able to discern the retreat of the water directly. Even though the view from the window of the ark would have revealed only water at this point, as it had for many months previously, contact with terra firma had been re-established, providing a fixed reference point for depth measurements. When the ark failed to lift off the mountain again, but rather proceeded slowly to settle in place, no doubt accompanied by some tilting of the decks, it would have been unequivocally clear to Noah that the Flood was waning and that the water was winning no longer.

The Flood Subsides

Noah observed that the Flood retreated steadily from the grounding of the ark onward (Genesis 8:5).

On 600/10/01 the tops of neighboring mountains became visible for the first time. The impression from Genesis 8:5 is that this was due to the Flood decreasing sufficiently for the tops of neighboring mountains

[7]Jack Finegan, *Handbook of Biblical Chronology* (Princeton: Princeton University Press, 1964), 32.

finally to poke out above the surface of the Flood water, like the Pacific islands which poke out above the surface of the Pacific Ocean today.

Forty days later, Noah began a series of bird experiments. The purpose of these experiments was clearly to ascertain living conditions outside the ark, as noted by many commentators.

Neither a date nor a day count is given in relation to the sending out of the first dove. However, Genesis 8:10 says in relation to the sending of the dove the second time, "So he waited yet another seven days; and again he sent out the dove from the ark." This seems to imply that there was a seven day interval between the sending of the raven and the sending of the dove the first time. I have assumed this is the case in the time chart of Figure 16.3.

Before moving on, notice that the date reference, 601/01/01, on which the covering of the ark was removed, probably does not refer to Noah's six hundred and first birthday. Rather, in the calendar assumed here, it means merely the day in which the first new moon of Noah's six hundred and first year was observed. Noah could have turned 601 at any time during the preceding lunar month.

Disembarking

A fascinating aspect of the chronology of the Flood, belying the prevalent modern notion that the Flood narrative is myth, is Noah's obvious timidity about leaving the ark. This is not explicitly mentioned in the narrative, but it comes through pretty clearly when the chronology of the Flood is considered. Figure 16.3 shows that Noah seems to have stayed aboard the ark more than two months longer than necessary.

Conditions inside the ark were very likely not all that pleasant. The ark is often pictured as something of a happy, floating zoo in children's storybooks, but a floating barn would probably be a more accurate image. Practical considerations suggest that the ark was probably a rather smelly place soon after the animals had boarded, and that it only got worse with time. Yet, even though Noah learned from his third experiment with the dove that conditions were livable outside the ark, he still waited another twenty-seven days before venturing to remove the covering of the ark on 601/01/01 to get some fresh air and have a good look around outside. And even though, having done so, he could now plainly see that the earth was dry all around, he still remained in the ark another fifty-five days—nearly two months—until God Himself commanded him to leave.

The implication here seems clear. Obviously, apart from a special revelation from God, Noah had no way of knowing whether the Flood was over. In the absence of such a revelation, he was clearly fearful that the Flood, which had swallowed the world so quickly and completely once, might suddenly return to do so again. His strategy seems to have been "better safe than sorry," even if it meant having to cope with a considerable degree of unpleasantness.

And in due time, on 601/02/27, the special revelation Noah needed was given, and Noah, his family, and the animals disembarked at long last. The ark had done its job, sheltering them successfully for 365 days. The year of God's judgment was now past, and they had God's promise that they would never again experience another year like it:[8]

> While the earth remains,
> Seedtime and harvest,
> And cold and heat,
> And summer and winter,
> And day and night
> Shall not cease.

[8]Genesis 8:22; NASB 1975.

Chapter Seventeen

A Matter of Interpretation

It is possible to adopt two different perspectives when reading the account of the Flood found in Genesis 7 and 8. One perspective is that these chapters record God's omniscient, omnipresent observations of the Flood. The other perspective is that these chapters record Noah's accurate but finite observations of the Flood.

These different perspectives affect interpretation of the narrative, especially in regard to the geographical extent of the Flood. If God's perspective is assumed, then many of the phenomena mentioned in these chapters appear to be global. If, however, Noah's perspective is adopted, then it is found that the question of the geographical extent of many of the phenomena mentioned in Genesis 7 and 8 cannot be settled with certainty from the biblical text alone.

The text says, for example, that "the rain fell upon the earth for forty days and forty nights."[1] If God's perspective is assumed, then this will mean that it rained everywhere on the globe without stopping for forty days and nights. This would go on to imply, for example, that the normally cold polar regions of the globe must have been warm during the Flood, so that rain would fall there rather than snow. But if Noah's perspective is assumed, then the forty days and nights of rain is an observation which is local to Noah, and hence does not demand forty days and nights of rain globally or warm polar temperatures.

As a second example, the text says that "all the high mountains everywhere under the heavens were covered."[2] If this is taken to be an

[1]Genesis 7:12; NASB 1975.
[2]Genesis 7:19; NASB 1975.

observation from God's perspective, then the Flood appears necessarily global. But if it is taken as an observation from Noah's perspective, then all that can be concluded from it is that all of the mountains within Noah's visible range (i.e., out to Noah's horizon in all directions) were submerged. This observation does not demand global coverage, as I will demonstrate in a subsequent chapter.

Which is the proper perspective?

There appears to be no way to settle this hermeneutical question on the basis of the biblical text alone. I will return to a discussion of it in Chapter 25 in relation to the geographical extent of the Flood waters. For now I wish only to make explicit the fact that I am personally convinced that it is Noah's perspective which is the proper choice, and to alert the reader that this interpretive perspective underlies and permeates the following chapters. My reasons for choosing this perspective come both from my study of the biblical narrative over several decades, and from my study of much extra-biblical data pertaining to the Flood, much of which will be shared in subsequent chapters.

As far as the biblical textual evidences are concerned, I have come to see Noah as somewhat of an early scientist. His ability to construct such a vessel as the ark, his ability to care for the many different types of animals aboard the ark, and his design and execution of the bird experiments all contribute to this impression. As a scientist myself, I find it difficult to escape the feeling that I am simply reading observations which Noah jotted down in his science notebook through much of Genesis 7 and 8, as I have previously mentioned.

The referencing of calendar dates to Noah's own birthday adds further to the impression that these are, in fact, Noah's personal observations of the Flood. The chronicling of events relative to one's birthday is still normal to much of common speech today, even with the ubiquitous availability of our advanced calendar system. For example, you would not be surprised to hear somebody say something like, "I graduated on the twenty-seventh day of June when I was eighteen, and got married on the ninth of July when I was twenty-three." Similarly, it feels we are hearing Noah say, "The tops of neighboring mountains became visible from the ark on the first day of the tenth month when I was six hundred."

An important consequence of this interpretive perspective is that the chronology of the Flood, shown in the time chart of Figure 16.3, cannot be applied simultaneously to every point on the surface of the globe. This is understood to be the chronology of *Noah's* experience of the Flood, not a universal chronology of the Flood equally applicable to the whole globe.

For example, the observation that the tops of the mountains became visible on 600/10/01 does not mean that mountains would have been first observed on that day no matter where an observer might have been situated on the earth. Similarly, the forty days and nights of rain must be regarded as a local, rather than a global observation. This by no means precludes the possibility of forty days and nights of rain elsewhere on the globe or even over the entire globe. What Noah observed locally may have been part of a global phenomenon, but it appears to me to be an interpretive error to jump to this conclusion on the basis of the biblical account of the Flood alone. Thus it is this more cautious interpretation which will be followed in subsequent chapters.

Chapter Eighteen

Conclusion to Part 2

> And just as it happened in the days of Noah, so it shall be also in the days of the Son of Man: they were eating, they were drinking, they were marrying, they were being given in marriage, until the day that Noah entered the ark, and the flood came and destroyed them all.[1]

The story of Noah's Flood found in Genesis gives every impression of being carefully recorded eyewitness observations. In concert with what is learned by experience, early in life, to be the hallmark of true historical narrative, the biblical narrative of Noah's Flood unfolds within a detailed calendrical framework. The events described are accompanied by a careful chronology, and this chronology displays several evidences of having originated, not in someone's imagination, but rather in real-life experiences.

While claims which are made by the narrative—the water covering high mountains for months, for example—are so extraordinary as to seem all but impossible, the context of these claims is everywhere sober narrative. The story, as it is told, does not have the flavor of legend. Rather, it feels like a personal diary, with a flavor much more akin to historical narrative than to legend.

This narrative, for all its terseness, is rich with information regarding the extraordinary historical event it records. Our task is neither to "tame" the ancient narrative, nor to extract its imagined "kernel of

[1]Luke 17:26–27; NASB 1975.

truth," nor to quarantine its real-life implications for us today by elevating it to a plane of "divine" fairy-tale. Our task is to come to grips with what we are plainly being told. A flood of unimaginable proportions nearly exterminated humanity 3520 B.C. Can we wake up to this fact? Can we register its numerous unsettling implications for humanity today?

Just how big was this flood? How can such a flood have happened? What was its driving mechanism? Could such a thing happen again? Let us press on.

PART 3

THE GEOGRAPHICAL EXTENT OF NOAH'S FLOOD

Chapter Nineteen

Noah's Flood at Devon Island

The Flood has been traced, by means of archaeological data, from ancient Mesopotamia (southeastern Iraq) to ancient Palestine (modern Israel) and then out to ancient Ireland. Can it be traced any farther around the globe?

Indeed, it can.

Ice Sheets as Flood Detectors

Vast ice sheets, some over a mile thick, cover many land areas in the polar regions of the globe today. These are the result of the buildup of annual snowfall over thousands of years. Of prominence in the north is the enormous Greenland ice sheet. In the south, covering nearly the entire continent of Antarctica, is the even larger Antarctic ice sheet.

Several ice sheets have been cored through to bedrock in modern times. The ice cores which have been retrieved have been intensely studied. They display thousands of years of annual layering in their upper portions, assisting in the task of establishing chronologies for these ice sheets. Simple layer counting techniques reveal that the oldest ice in these ice sheets predates Noah's Flood by many millennia.

If the Flood covered an ice sheet, then one of two things would have had to happen. Either the ice sheet would have remained submerged beneath the Flood water, or it would have broken away from its bed and floated like a gigantic iceberg in the Flood water.

If an ice sheet had remained submerged, meltback of the surface of the ice sheet by the action of the Flood waters would have occurred. If it had floated, meltback of the underside of the ice sheet would have occurred.

Thus, there is every reason to expect that if the Flood extended to these ice sheets, then its effects should yet be detectable in the ice cores which have been retrieved from them.

Glaciology and the Flood

Melting

Ice at atmospheric pressure melts at 0°C. At 300 atmospheres pressure, such as would be encountered under a two-mile water column, ice melts at −2.32°C.[1] Meanwhile, deep ocean water, which makes up most of the bulk of the oceans today, is typically within a few degrees of 4°C, and surface water temperatures at low latitudes exceed 20°C.[2]

Thus the Flood is expected to have melted the ice sheets wherever it came in contact with them, as stated above. The rate of melting would have depended on the actual temperature of the Flood water at a given time and location, in addition to other factors, of course, but melting is expected in any event.

Top or Bottom

Ice floats in water. Even the highly pressurized ice at the bottom of a large ice sheet will float, since it is less dense than water.

But many ice sheets are observed to be frozen to their beds at present. Whether an ice sheet would have floated and melted on its underside during the Flood, or remained under water and melted on its topside, depends on whether the buoyant force on that particular ice sheet was strong enough to overcome the tensile strength of the ice holding it to its bed.

To Float or Not to Float

Whether an ice sheet will float or not in deep enough ocean water can be calculated using the fundamental laws of physics and the measured

---

[1] *CRC Handbook of Chemistry and Physics* 77th edition, section 6, page 15.

[2] George L. Pickard and William J. Emery, *Descriptive Physical Oceanography: an Introduction*, 4th (SI) enlarged edition (New York: Pergamon Press, 1982), 34–43.

properties of glacier ice. If the force of buoyancy acting on an ice sheet due to the presence of water is greater than the weight of the ice sheet plus the force due to the ultimate tensile strength of the ice holding it to its bed, then the ice will break away from the bed and float. Otherwise it will stay submerged.

In Appendix A (page 317) it is shown that an ice column having a top-to-bottom thickness less than 2200±600 meters will, when submerged, remain submerged, attached to its bed.

The uncertainty in this calculated thickness is, unfortunately, large. This hampers ability to make clear-cut predictions for ice sheets having average thicknesses near 2200 meters.

Applying the usual rules, from the branch of mathematics known as statistics, to this calculated critical height and its uncertainty indicates that there is about a two-in-three chance the true critical height for ice sheets lies somewhere between 1600 and 2800 meters. Conversely, there is about a one-in-three chance it lies outside this range.

Unfortunately, the thicknesses of the largest (and most intensely studied) polar ice sheets—the Greenland and Antarctic ice sheets—appear to be comparable to this calculated critical height range. For example, cores of lengths 1220, 1390, 2620, 2950, and 3050 meters have been drilled at various points on the Greenland ice sheet. This makes prediction of just what to expect the Flood to have done to these large ice sheets relatively difficult. Would the thicker sections of the ice sheet have provided enough buoyancy to lift the entire sheet, or would the thinner sections have kept the entire sheet "glued" down? Or would the sheet have fractured into several pieces, with some pieces remaining submerged while others floated?

Fortunately, ice cores have also been taken from a few smaller ice sheets, and with these the expected action of the Flood is unambiguous.

The Devon Island Ice Cap

The most intensely studied smaller ice sheet I have been able to find, suitable to the present purpose of determining the extent of the Flood, is an ice cap on Devon Island in Northern Canada (Figure 19.1). The suitability of this smaller ice sheet results from the chronological data which have been gathered on it, and from the fact that its oldest ice dates much further back than 5500 years ago, when the Flood occurred.

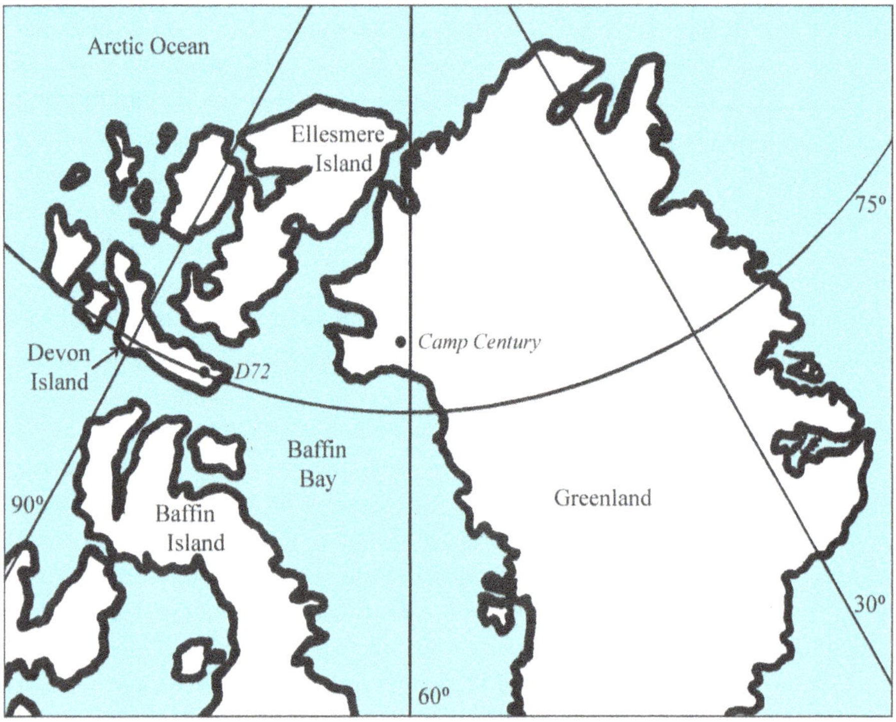

Figure 19.1: Location of Devon Island, west of Greenland.

This "smaller" ice cap on Devon Island is still quite massive. It covers an area of nearly four million acres (15,600 km^2).[3]

Two bore holes, D72 and D73, were drilled through the Devon Island ice cap to bedrock. D72, drilled in 1972, was 298.9 meters deep. D73 was drilled in 1973 and was 299.4 meters deep. Paterson reports that "measured ice thicknesses are in the range 200 to 1,000 m."[4] Thus, an average thickness much less than the critical height of 2200±600 meters calculated above is indicated, and it may be confidently asserted that the Devon Island ice cap would have remained frozen to its bed throughout the Flood. Consequently, if the Flood extended to Devon Island to sufficient depth to cover the pre-Flood Devon Island ice cap, then significant

[3] W. S. B. Paterson, "Vertical Strain-rate Measurements in an Arctic Ice Cap and Deductions from Them," *Journal of Glaciology* 17.75 (1976): 4.

[4] W. S. B. Paterson, "Vertical Strain-rate Measurements in an Arctic Ice Cap and Deductions from Them," *Journal of Glaciology* 17.75 (1976): 4.

meltback of the top side of the ice cap must have occurred.

Is there any evidence of topside meltback of the Devon Island ice cap 5500 years ago?

Indeed there is.

Evidence of Topside Meltback at Devon Island

Perhaps the most direct route to seeing this, for the non-glaciologist, is via a result published in the *Journal of Glaciology* by N. Reeh and W.S.B. Paterson regarding the Devon Island ice cap following an extensive effort to model the properties of the ice cap numerically. They found that the ice cap seems to have been thickening from an unexpectedly thin state some 5000 to 6000 years ago:[5]

> Application of a simplified non-steady-state flow model ... indicates that the measured [Devon Island ice cap annual] layer thicknesses are compatible with a gradual thickening of the ice cap near the bore hole [D72] from about 100 m [100 meters] 5000 – 6000 years ago to the present thickness of about 300 m. This compares well with the 250 m of thickening needed to reconcile the Devon Island and Camp Century [Greenland] oxygen-isotope profiles over the past 5000 a [5000 years], ... These ice-thickness changes are surprisingly large.

The observation that the ice cap was unexpectedly thin 5000 to 6000 years ago and that it has thickened considerably since that time does not tell us whether the ice cap had come to this thin state by loss of ice from its top or its bottom side. But this question is settled by further observations on the ice cores, D72 and D73. From study of these cores glaciologists Koerner and Fisher concluded, "... the ice has almost certainly been frozen to its bed throughout its history."[6] Thus any meltback of the ice cap resulting in the observed thinness 5000 to 6000 years ago must have occurred from its top side, not its underside.

---

[5]N. Reeh and W. S. B. Paterson, "Application of a Flow Model to the Ice-divide Region of Devon Island Ice Cap, Canada," *Journal of Glaciology* 34.116 (1988): 62.

[6]R. M. Koerner and D. A. Fisher, "Discontinuous Flow, Ice Texture, and Dirt Content in the Basal Layers of the Devon Island Ice Cap," *Journal of Glaciology* 23.89 (1979): 218.

Elaboration

A Discontinuity

The idea that the top of the Devon Island ice cap was melted back significantly by the Flood suggests the possibility of a discernible discontinuity of some sort between old, pre-Flood ice and new, post-Flood ice in the ice cores.

As it turns out, a discontinuity in the comparison of measured oxygen isotope ratios in the two cores shows up 13 meters above the base of the Devon Island cores. Paterson *et al.* report that:[7]

> The [oxygen] isotope profiles from the two boreholes closely resemble each other except near the base of the ice. The correlation between 50-yr mean δ [oxygen isotope ratio] values is 0.965 between the surface and 13 m above the bed, but only 0.449 between 13 and 5 m.

The two cores were drilled only 27 meters apart. Effort was made to situate them on the same ice flow line. Thus their oxygen isotope ratios should closely resemble one another. The fact that they do not do so below 13 meters above bedrock is clear evidence of some sort of disturbance.

The original researchers suggested that the observed loss of correlation may be due to unevenness of the bedrock over which the ice slowly moves. But they noted that the ice from the cores fails to show the sorts of crystallographic evidence expected to support this suggestion.

While some of the loss of correlation in the bottom ice may be due to perturbations in ice flow over an uneven bed, especially in the lowest few meters, I suggest the principal reason the sections from the two cores between 13 and 5 meters show poor correlation is that the ice below 13 meters is thousands of years pre-Flood in origin, while the ice above 13 meters is post-Flood in origin. If pre-Flood ice flow lines were different from presently observed flow lines, then the lower sections of the two cores may have originated from quite different places on the surface of the pre-Flood ice sheet. In that case, the overall trends in oxygen isotope ratios would be expected to be similar (as is observed), but detailed correlation

[7]W. S. B. Paterson, R. M. Koerner, D. Fisher, S. J. Johnsen, H. B. Clausen, W. Dansgaard, P. Bucher, and H. Oeschger, "An Oxygen-isotope Climatic Record from the Devon Island Ice Cap, Arctic Canada," *Nature* 266 (7 April 1977): 509.

would not be expected. In addition, the residual pre-Flood ice would probably have been melted unevenly by the action of the Flood, perhaps even with melting and refreezing extending along fissures into the old ice. All such Flood-induced irregularities would contribute to a lack of correlation between features from neighboring cores in the sections of the cores pre-dating the Flood.

The Date

Most important to the idea that this discontinuity is due to the Flood having melted back the top side of the Devon Island ice cap is the date of the observed discontinuity. Biblical chronology says the Flood happened 3520±21 B.C. Does the ice immediately above 13 meters date to within ice core dating uncertainties of 3520±21 B.C.?

Yes, it does.

The date of ice at the depth of interest can be determined in a fairly straightforward manner from the stratigraphy of the ice itself. Appendix B (page 321) shows the details of my calculation of the date of the ice immediately above 13 meters. I found 3430 B.C. This is just 90 years shy of the biblical chronology date for the Flood.

While it is difficult to assess quantitatively the uncertainty in this calculated ice core date arising from the interpolations and extrapolation used to obtain it (Appendix B), it must certainly be greater than a century. A century corresponds to less than a 2% measurement error. Experience teaches that 10% uncertainty is generally a reasonable rule of thumb, and this corresponds to ±540 years in the present case.

Thus, the date of the oldest ice above the 13 meter-above-bedrock ice is found to be well within normal ice core dating uncertainties of the date for the Flood, supporting the claim that the Devon Island ice cap was melted back on its top side by the Flood.

Failure of the Steady State Assumption

Further evidence of topside meltback of the Devon Island ice cap is provided by its failure to satisfy the usual steady state assumption for modeling existing ice sheets.

In steady state, the properties of an ice sheet—its overall thickness and the thickness of its annual layers at a given height above bedrock, for example—remain constant with respect to time, even though new ice is being added to the top side of the ice sheet year by year. This results from the fact that old ice is squeezed out the sides of the ice cap, by the weight of the overlying ice and snow, at the same rate as new snow accumulates on top each year.

Conventional scientific thinking—which presently makes no allowance for Noah's Flood, as I have previously discussed—expects ice sheets to be in steady state today. This is because climate-controlled environmental conditions affecting these ice sheets are expected to have been more or less uniform since the last glacial period ended over eleven thousand years ago. The accumulation of snow on Devon Island, though fluctuating from year to year, is expected to have been approximately constant on average, with no significant long term trend, since the last glacial period. The temperature, too, is expected to have been roughly constant on average. And since nothing is known to secular science which might have upset the simple year-by-year accumulation and loss of ice at Devon Island, conventional thinking says that basic properties of the Devon Island ice cap should be able to be successfully explained and mathematically modeled assuming steady state conditions within the ice cap today.

This expectation contrasts sharply with what is to be expected if Noah's Flood reached Devon Island and submerged its ice cap for several months. In that case, meltback of the upper surface of the ice cap would have taken place 5500 years ago. Such a one-time meltback severely contradicts the steady state assumption of roughly constant loss of old ice from the ice cap year by year. Thus, the presence of the Flood at Devon Island would be revealed by failure of the steady state assumption in any effort to model the properties of the ice cap numerically.

Paterson and Waddington applied a steady state model to the Devon Island ice core data in 1984 and found, much to their surprise, that it didn't work. They concluded "... the steady state assumption has broken down."[8] Reeh and Paterson tried again four years later, with a more detailed steady state model. They found, once again, that the model failed to agree with the measured properties of the ice cap. To obtain reasonable agreement required "application of a simplified non-steady-state flow model."[9]

---

[8]W. S. B. Paterson and E. D. Waddington, "Past Precipitation Rates Derived from ice core Measurements: Methods and Data Analysis," *Reviews of Geophysics and Space Physics* 22.2 (May 1984): 126.

[9]N. Reeh and W. S. B. Paterson, "Application of a Flow Model to the Ice-divide

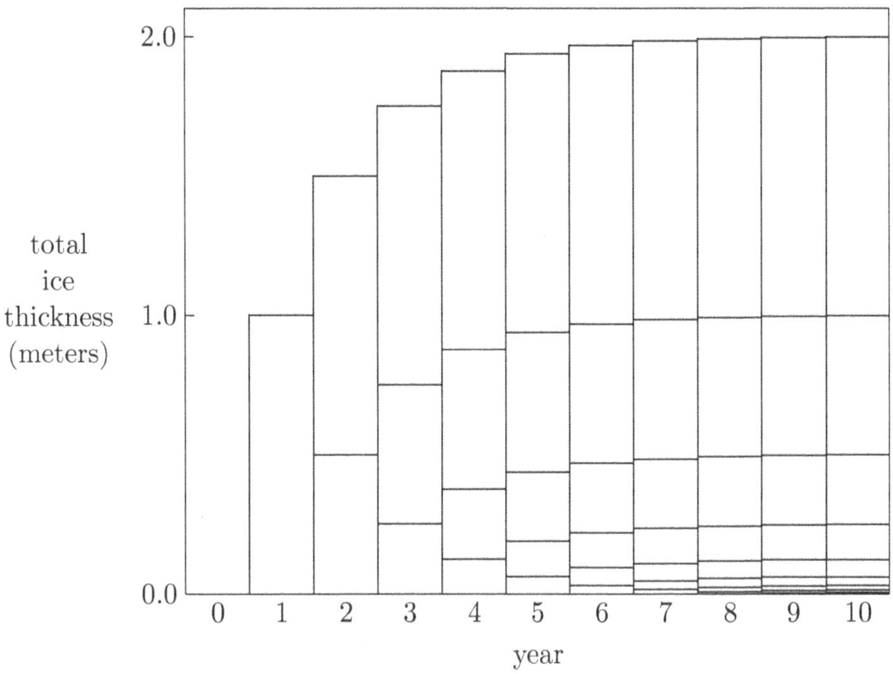

Figure 19.2: Simplified illustration of the growth of an ice sheet showing how bottom layers thin while upper layers retain their same thickness as the ice sheet ages.

Too Thick Annual Layers

When an ice sheet loses ice due to meltback, measured annual layer thicknesses will be found to be increasingly too thick with depth relative to steady state expectations.

To see this, consider the time development of an ice sheet following a meltback event (Figure 19.2). For the sake of simplicity, assume the ice sheet receives just one meter of snow each year. Further assume that in year 0 the ice sheet has been melted back to bedrock. (The depth and time scale are unimportant here; it is only the principles which are important.) In the first year after the meltback, the snow will accumulate to one meter depth above bedrock. In the second year, another meter of snow will accumulate. This will weigh down the first year's accumulation and compress it somewhat. For the sake of simplicity once again, assume

Region of Devon Island Ice Cap, Canada," *Journal of Glaciology* 34.116 (1988): 62.

that each layer compresses each year to one half the thickness it had the year before due to the weight of the overlying snow. (In real life thinning occurs by compression near the top, giving way to horizontal flow deeper in the ice, but such details are unimportant here.) If this is allowed to go on indefinitely, simple mathematics shows that this hypothetical "ice sheet" will achieve a steady state thickness of two meters.

Figure 19.2 (page 141) shows the approach to steady state for the first ten years in this case. Compare the ice sheet layer thicknesses after four years with the ice sheet layers after ten years. Notice that the upper layer thicknesses are identical—the only difference is that they are at a lower altitude at four years than they are at ten years. In contrast to this, however, the bottom layers in the four-year-old ice sheet are much thicker than the bottom layers in the ten-year-old ice sheet.

As an ice sheet ages toward steady state following a meltback event, its top layers show no change in thickness, but its bottom layers grow thinner and thinner. If steady state is mistakenly assumed in an ice sheet which is growing and has not yet come to steady state, it will be found that the upper layers are the expected thicknesses, but the lower layers are increasingly too thick for their depth.

Paterson and Waddington were the first to observe this effect in the Devon Island ice cap.[10] Their observation resulted when they applied a steady state model, mentioned above, to the ice cap.

Reeh and Paterson's more sophisticated steady state model, also mentioned above, subsequently arrived at the same result:[11]

> Calculated layer thicknesses are compared with the measured ones... Down to a depth of 136 m, where the ice is about 900 years old, calculated values are within 10% of the measurements, except for one point. Below this, the measured thicknesses exceed the calculated ones and their ratio increases with depth. At 267 m, for example, the measured layer thickness is six times the calculated one.

[10]W. S. B. Paterson and E. D. Waddington, "Past Precipitation Rates Derived from ice core Measurements: Methods and Data Analysis," *Reviews of Geophysics and Space Physics* 22.2 (May 1984): 126.

[11]N. Reeh and W. S. B. Paterson, "Application of a Flow Model to the Ice-divide Region of Devon Island Ice Cap, Canada," *Journal of Glaciology* 34.116 (1988): 62.

Increasing Altitude

Figure 19.2 (page 141) illustrates another feature of an ice cap following meltback. The altitude of the top side of the ice cap will increase as the ice sheet rebounds toward steady state as a result of fresh snow deposition year by year following the meltback episode. Notice the increase in height of the hypothetical ice sheet from year 0 to year 10 in Figure 19.2. It is rapid at first, then levels off.

The Devon Island ice cap displays this feature also. Its increasing altitude following meltback is revealed by oxygen isotope measurements made on the ice of the cores taken from the ice sheet (Figure 19.3, page 144).[12]

Oxygen isotope analysis is a standard tool of science used more or less routinely in the study of ice cores today. The concentrations of isotopes of oxygen (oxygen atoms of differing masses) are measured in the ice at various heights above bedrock in the ice core. At each measurement point, the per mil (i.e., parts per thousand) difference from a standard sample of the ratio of heavy to light oxygen is expressed as δO^{18} (pronounced "delta oh eighteen").

A graph of δO^{18} versus height above bedrock reveals information regarding the deposition altitude of snow on an ice sheet. Deposition of snow at a higher elevation results in a lower δO^{18} value. For an ice sheet melted back suddenly at some point in time, δO^{18} should show a step from lower to higher values at that point in time, followed by a slow return to lower values as the ice sheet rebounded to its normal height in subsequent years.

Unfortunately, at least as far as simplicity of data interpretation is concerned, altitude of snow deposition is not the only factor affecting δO^{18}. A change in the climate affecting mean air temperature may also be expected to affect δO^{18}, as may changing patterns of atmospheric circulation, for example. But other measured parameters, such as the annual layer thicknesses just discussed, and comparison with ice cores from other nearby ice sheets, such as the Greenland ice sheet in the present case,[13] may be used to clarify the cause of an observed change in

---

[12]The data for this figure are from W. S. B. Paterson, R. M. Koerner, D. Fisher, S. J. Johnsen, H. B. Clausen, W. Dansgaard, P. Bucher, and H. Oeschger, "An Oxygen-isotope Climatic Record from the Devon Island Ice Cap, Arctic Canada," *Nature* 266 (7 April 1977): 510.

[13]W. S. B. Paterson, R. M. Koerner, D. Fisher, S. J. Johnsen, H. B. Clausen, W.

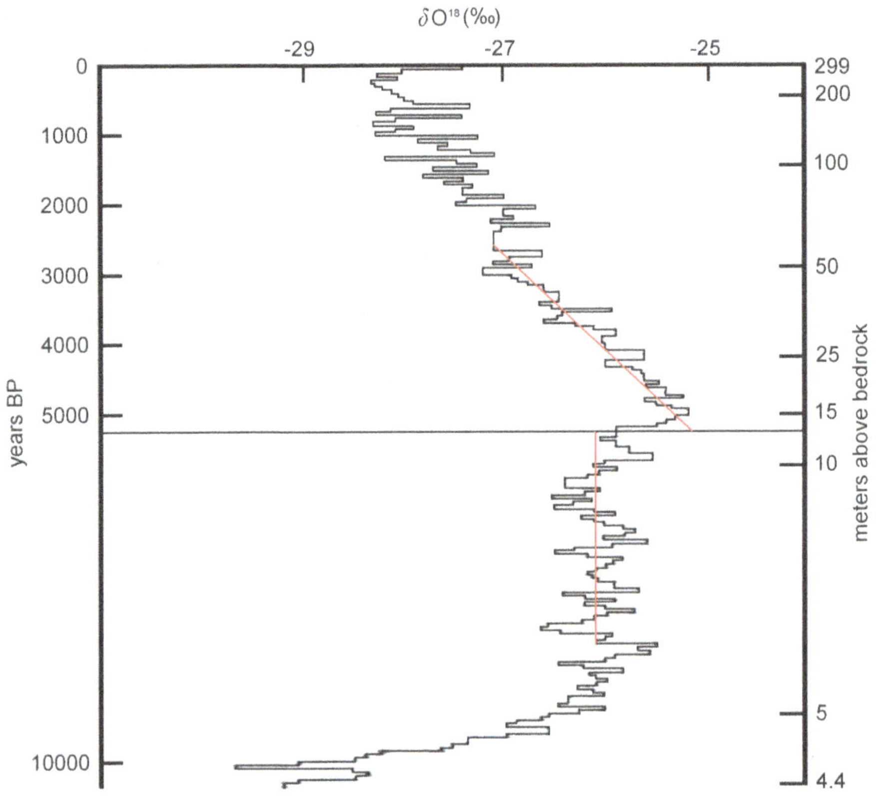

Figure 19.3: Oxygen isotope ratios measured versus height above bedrock in the Devon Island ice cap. The time scale shown to the left of the graph is that of the original researchers. This time scale was used to group the raw data from both cores D72 and D73 into 50-year means. The red lines are drawn by me, by eye, to show the approximate long-term trends before and after the 13 meter-above-bedrock discontinuity. (Data are from Paterson *et al.* See footnote 12 for the complete reference.)

δO^{18}. And when this is done it is clear that most of the observed change must be attributed to a change in altitude in the case of the Devon Island ice cap.

Figure 19.3 thus displays the thinning of the ice cap at the close of the last glacial period roughly 11,500 years ago. A steady state thickness seems to have been attained soon thereafter. I have drawn the lower red line in the figure to represent this apparent longterm steady-state pre-Flood ice thickness. A step down to lower altitude is then apparent, moving from the pre-Flood (below 13 meters) ice to the post-Flood (above 13 meters) ice. This is most clearly seen by considering the longterm trend in the data once again, which I have represented by the upper red line. The upper part of the graph shows continued thickening of the ice sheet toward a new steady state thickness.

Conclusion

It is possible to estimate how much the Devon Island ice cap was melted back 5500 years ago using the Figure 19.3 δO^{18} data. After comparing the Devon Island δO^{18} data to δO^{18} data from Camp Century, Greenland, just across Baffin Bay from Devon Island, Paterson *et al.* observed, "the additional 1.6‰ at Devon Island would correspond to a thickness change of about 250 m."[14] Since the ice cap is currently 300 meters thick at the drill site, this suggests that only about 50 meters of the original ice cap remained near the borehole following meltback 5500 years ago. (This original 50 meters has thinned to just 13 meters today due to the weight of the overlying ice.) The step to higher δO^{18} between the red lines at 13 meters in Figure 19.3 represents a reduction in altitude of the ice sheet of about 140 meters near the borehole. So the original thickness appears to have been roughly 190 meters.

Thus the data from Devon Island paint a picture of an ice cap frozen to its bed, suddenly melted back from the top about 5500 years ago, reducing it from its original thickness of about 190 meters near the bore-

---

Dansgaard, P. Bucher, and H. Oeschger, "An Oxygen-isotope Climatic Record from the Devon Island Ice Cap, Arctic Canada," *Nature* 266 (7 April 1977): 508–511.

[14]W. S. B. Paterson, R. M. Koerner, D. Fisher, S. J. Johnsen, H. B. Clausen, W. Dansgaard, P. Bucher, and H. Oeschger, "An Oxygen-isotope Climatic Record from the Devon Island Ice Cap, Arctic Canada," *Nature* 266 (7 April 1977): 510.

hole to a mere 50 meters, followed by resumption of growth toward a new steady state balance.

This picture is most unexpected within the presently established secular view of earth history. My search of the technical literature failed to turn up any cogent explanation of it, despite some researchers obviously having wrestled seriously with the problem. In sharp contrast, this picture is both anticipated and explained by Noah's Flood having extended to Devon Island.

Can Noah's Flood be traced any farther around the globe than from Mesopotamia to Palestine and Ireland? The answer is clearly yes. Noah's Flood extended also to Devon Island, Canada.

Chapter Twenty

Objection: Noah's Flood is Missing at Ellesmere Island

Ellesmere Island, Canada, lies north of Devon Island (Figure 20.1). Agas-

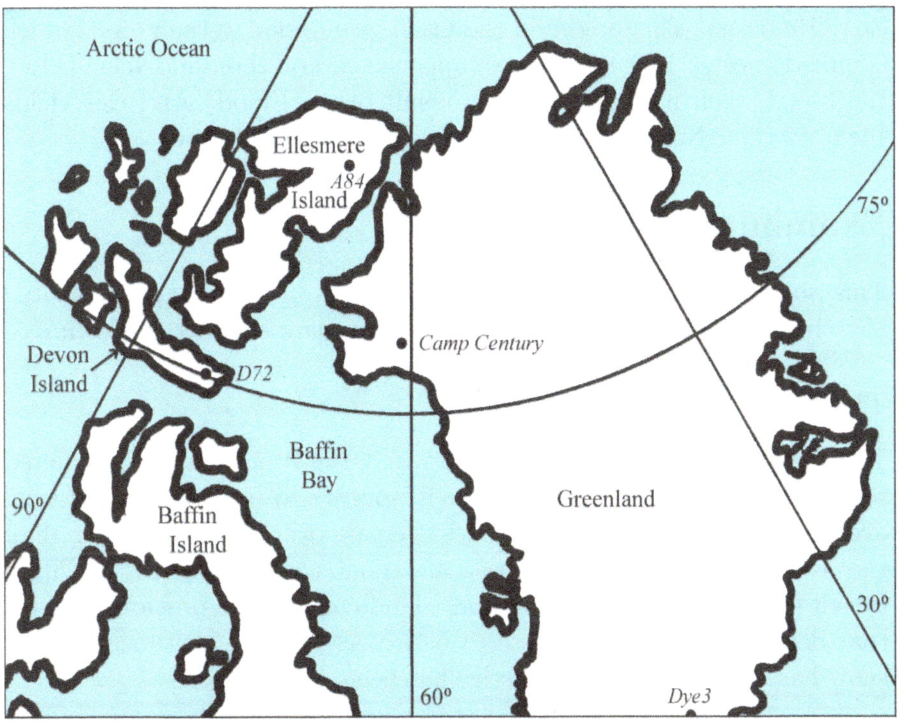

Figure 20.1: Location of bore hole A84 on Ellesmere Island.

siz ice cap on Ellesmere Island was cored through to bedrock in 1977, 1979, 1984, and again in 1987, yielding ice cores A77, A79, A84, and A87 respectively. The present chapter focuses on A84 since it is "considered to be the most reliable of the suite" by the original researchers.[1]

The Objection

Noah's Flood, it has now been shown, extended from southeastern Iraq to Devon Island. This is a distance of roughly 7250 kilometers. From the Devon Island ice cap to the Agassiz ice cap on Ellesmere Island is only about one tenth of this distance. Since the Flood covered high mountains for months, and since it covered the Devon Island ice cap deep enough and long enough to melt it back some 140 meters, it surely must also have extended to the Agassiz ice cap on Ellesmere Island and melted it back too.

Is there any evidence of topside meltback of the Agassiz ice cap? Indeed there is, but the original researchers date this meltback to the early Holocene. This is several thousand years prior to the Flood, which is mid-Holocene. If the meltback happened several thousand years before the Flood, then it clearly is not a result of the Flood. And this yields the objection: Noah's Flood is missing at Ellesmere Island.

Examining the Objection

This objection rests upon the claim that the Agassiz ice cap was melted back in the early Holocene. Do the available data support this claim?

The Data

The fundamental evidence indicating that the Agassiz ice cap suffered serious meltback in the past is that it appears to have been thickening over much of its history. δO^{18} values decrease over much of the length of the A84 core: from about 15 to 95 meters above bedrock (Figure 20.2). Recall from the study of the Devon Island cores in the previous chapter that decreasing δO^{18} values are to be expected if the altitude at which snow has been deposited on the ice has been increasing.

[1]David A. Fisher, Roy M. Koerner, and Niels Reeh, "Holocene Climatic Records from Agassiz Ice Cap, Ellesmere Island, NWT, Canada," *The Holocene* 5.1 (1995): 20.

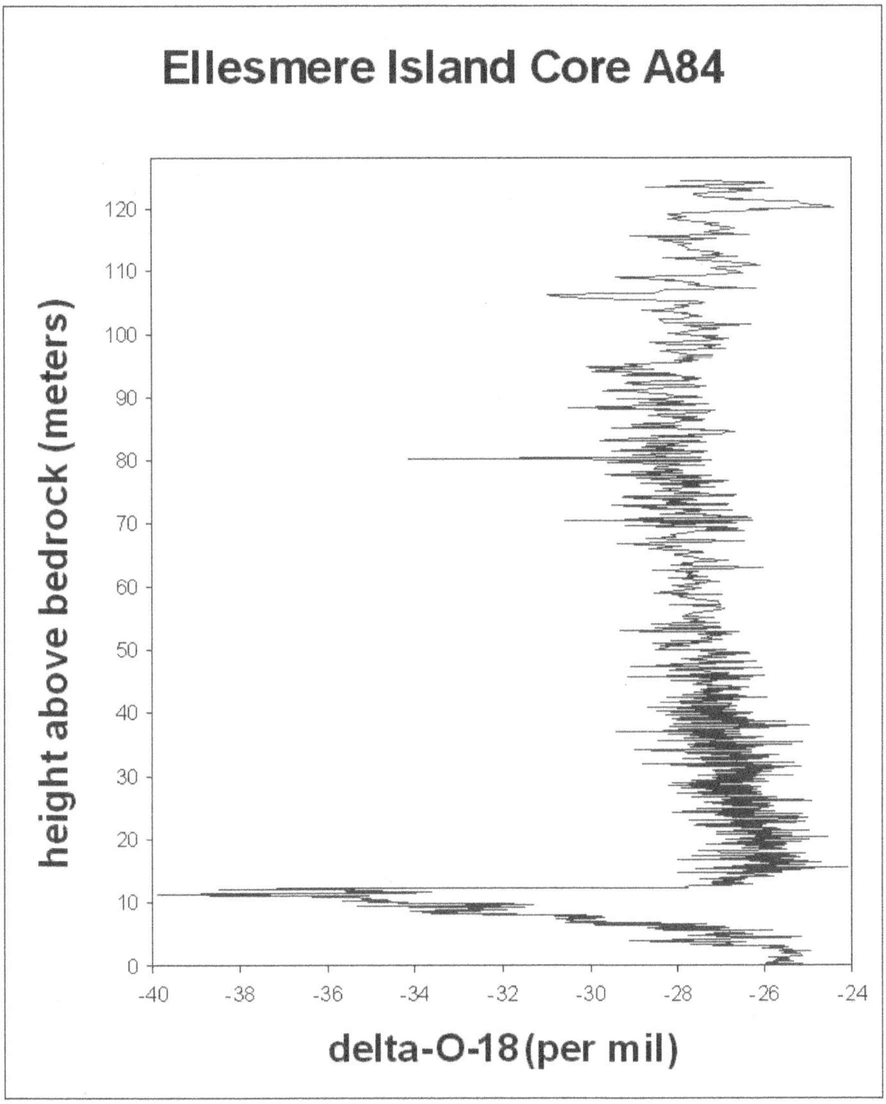

Figure 20.2: Oxygen isotope data for the A84 ice core from Ellesmere Island, Canada, from bedrock to nearly the top of the core at 127 meters.
(Data were obtained via http://www.ncdc.noaa.gov/paleo/icecore/polar/ Agassiz/raw.html. The graph combines all data from the files: mer84-1u_a84.txt, mer84-5_a84.txt, mer84-25_a84.txt and mer84-ws_a84.txt.)

The original researchers notice that the ice cap appears to have been thickening, but they do not, of course, see this thickening as a consequence of meltback by Noah's Flood 5500 years ago. To modern science Noah's Flood is but a myth, and myths don't melt real ice. Rather they picture loss of ice due to a "period of extreme summer melt in the early Holocene"[2] lasting at least a thousand years, and they suggest that "some very early Holocene ice might be missing."[3] The dates they have in mind for the early Holocene are prior to about 8000 years ago.

It is immediately curious that two ice caps, Devon Island and Agassiz, located in such relative proximity to one another, should show evidence of unusual melting followed by prolonged thickening but at entirely different times. All agree that Devon Island ice cap was melted back 5500 years ago. Why would serious meltback, which appears to be a rare phenomenon, happen 5500 years ago on Devon Island and entirely independently over 8000 years ago on the neighboring Ellesmere Island? Koerner has observed, in regard to summer conditions over the past century, that the Devon Island ice cap is generally representative of all the ice caps in this region, including the Agassiz ice cap:[4]

> A comparison with the melt-layer ice percentage in cores from the other major Canadian Arctic ice caps shows that the variation of summer conditions found for the Devon Island ice cap is representative for all the large ice caps for about 90 percent of the time.

Wouldn't it make more sense for the ice which is missing from both ice caps to be missing at the same time?

But for this to be the case, the chronology which Fisher *et al.* have worked out for the Agassiz ice cap must be seriously in error.

Examining the Chronology of A84

It is not hard to see how the chronology of the Agassiz ice cap might have been gotten wrong. The most pronounced feature in the δO^{18} record is

---

[2]David A. Fisher, Roy M. Koerner, and Niels Reeh, "Holocene Climatic Records from Agassiz Ice Cap, Ellesmere Island, NWT, Canada," *The Holocene* 5.1 (1995): 21.

[3]David A. Fisher, Roy M. Koerner, and Niels Reeh, "Holocene Climatic Records from Agassiz Ice Cap, Ellesmere Island, NWT, Canada," *The Holocene* 5.1 (1995): 23.

[4]R. M. Koerner, "Devon Island Ice Cap: Stratigraphy and Paleoclimate," *Science* 196 (1 April 1977): 18.

the sudden large change in oxygen isotope ratio near 13 meters above bedrock (Figure 20.2, page 149). For a modern scientist, taught to assume more or less uniform conditions since the end of the last glacial period, there is really only one way to explain this sudden change. It must correspond to the close of the glacial period itself, when rising temperatures caused thinning and retreat of the vast glacial period ice sheets. There really is no other way to explain such a large, abrupt change within conventional modern scientific thought. No other mechanism is available to accomplish such a dramatic change. Thus, within conventional thinking, this sudden change is "known" to correspond to the end of the last glacial period.

This "known" then becomes a fixed point for the chronology of the core. This sudden change, it is assumed, fixes the date of the ice at this height in the core to the end of the last glacial period, variously dated to 10 to 12 thousand years ago.

> The large abrupt change in oxygen-isotope ratio near the bottom of each core provides another reference horizon. This "step," which coincides with a sudden change in the concentration of microparticles (many more particles in the older ice) has also been found in the Devon Island and Camp Century cores and at Byrd Station, Dome C and Vostok in Antarctica. This horizon represents the end of the last glaciation; we choose the date given by the Camp Century time scale, 10,500 yr BP. . . [5]

> The transition from glacial period to Holocene as defined by the large sudden δ step is given the calendar age of 11 550 BP.[6]

Once the date of this sudden change has been assumed, the chronology of the ice core from the surface (of known modern date) to the sudden change is necessarily constrained to fit between these two fixed points, and any subjectivity in construction of the intervening chronology will

[5] D. A. Fisher, R. M. Koerner, W. S. B. Paterson, W. Dansgaard, N. Gundestrup, and N. Reeh, "Effect of Wind Scouring on Climatic Records from Ice-core Oxygen-isotope Profiles," *Nature* 301 (20 January 1983): 207–208.

[6] David A. Fisher, Roy M. Koerner, and Niels Reeh, "Holocene Climatic Records from Agassiz Ice Cap, Ellesmere Island, NWT, Canada," *The Holocene* 5.1 (1995): 21.

unavoidably be influenced toward producing an acceptable fit to these fixed end points. This is not much of a problem at the upper end of the core, where the date of the top of the core is truly known, and where there is a relatively large thickness of ice per year. But near the bottom of the core, it is asking for trouble. Near the bottom of the core the ice has been severely compressed and thinned as a result of the weight of the overlying ice pressing down for thousands of years. A millimeter offset in a few measurements near the bottom can easily result in dating offsets of thousands of years. And, unfortunately, the method used by the original researchers to construct a chronology for the A84 core involved considerable subjectivity.

Layer Counting Not Used

Normally, chronology building begins with measurements of annual layer thickness versus depth in the core, as Appendix B (page 321) illustrates with the Devon Island core, D72. This usually involves measuring the seasonally varying concentration of microparticles (i.e., dust) in the ice. This is already far from ideal, as far as objective measurement goes, since the individual making the measurements must decide in an intrinsically noisy signal which peaks and valleys delineate individual years. (Picture snow blowing around on the surface of the ice sheet each winter, and snow melting back by varying amounts and with varying frequency on the surface of the ice sheet each summer.) The researcher has subjective latitude ranging from "count everything possible" to "eliminate everything questionable." It is all but impossible for the researcher not to be influenced toward one end or the other of this range by his or her knowledge of how many years are needed to fill up the chronology between the fixed end points.

But the situation is worse even than this for A84. It appears that techniques involving even greater subjectivity were used instead of annual layer counting in the construction of its chronology. The original researchers explained, a year before the A84 core was obtained:[7]

> Because Hole 79 [A79] is only 139 m deep, the depth where layers become too thin to sample corresponds to much younger

[7]D. A. Fisher, R. M. Koerner, W. S. B. Paterson, W. Dansgaard, N. Gundestrup, and N. Reeh, "Effect of Wind Scouring on Climatic Records from Ice-core Oxygen-isotope Profiles," *Nature* 301 (20 January 1983): 207.

> ice than in Hole 77 [A77]. We therefore estimated ages from a flow model.

Since A84—just 128 meters deep—is shorter yet than A79, its initial time scale must also have been simply estimated from a flow model without benefit of direct layer counting.

Thus the actual process used to obtain a chronology for the A84 core appears to have been first to estimate the time scale using a theoretical flow model and then to adjust the time scale by matching volcanic acidity peaks in the ice core to known historical dates of volcanoes or to dates assigned to unknown volcanoes in another ice core:[8]

> The initial theoretical timescales are 'tuned' using volcanic acid-layer stratigraphy. The acid stratigraphy was detected using the ECM method and the dates were assigned using known historical records or the date given from the Dye 3 [Greenland] ice core.

Unfortunately, both the theoretical flow model and the process by which the timescales were "tuned" using acidity peaks are susceptible to inadvertent subjective bias seriously distorting the time scale, especially near the bottom of the core.

Flow Model

The theoretical flow model used for the A84 core likely involved steady state assumptions—a constant annual accumulation rate of new ice since the early Holocene, for example—just as the flow model used for the A77 core did.[9] But if the Flood significantly melted back the Agassiz ice cap, as was found to be the case with the Devon Island ice cap in the previous chapter, then such a flow model will be seriously in error for the lowest part of the core. The flow model will calculate annual layer thicknesses which are much too thin in the deeper parts of the core, just as was seen to be the case at Devon Island, and this means that it will overestimate the age of the deeper parts of the core.

---

[8]David A. Fisher, Roy M. Koerner, and Niels Reeh, "Holocene Climatic Records from Agassiz Ice Cap, Ellesmere Island, NWT, Canada," *The Holocene* 5.1 (1995): 20.

[9]D. A. Fisher, R. M. Koerner, W. S. B. Paterson, W. Dansgaard, N. Gundestrup, and N. Reeh, "Effect of Wind Scouring on Climatic Records from Ice-core Oxygen-isotope Profiles," *Nature* 301 (20 January 1983): 207.

The modelers will not likely suspect that their model has this problem, however, because they have been unable to measure the actual layer thicknesses at depth in core A84, and the date they have assigned to the sudden δ step transition is thousands of years earlier than the Flood, which encourages overestimation of the age of the ice back to that transition.

Acidity Peaks

The assignment of acidity peaks is then easily misled. Figure 20.3 shows the available acidity peak data for A84. Notice the large number of peaks present in these data.

Not all of these peaks are due to volcanic eruptions, by any means. Acids are naturally present in the atmosphere even in the absence of volcanoes. These acids deposit on the surface of the ice sheet just as volcanic acids do. And, being very water soluble, these acids can be carried down into the snow in percolation due to summer melting, which can produce acidity peaks.

The problem is how to distinguish between peaks due to percolation and peaks due to volcanoes.

One obvious approach might be to limit "tuning" to just the unusually large peaks, under the assumption that peaks which are well above the normal background "noise" have low probability of being due to percolation. But notice that all of the large peaks lie in roughly the top half of this ice core. This means that the bottom half of this ice core, which is in greatest need of absolute chronological control, lacks the sort of truly distinctive peaks which are needed to help eliminate subjective bias in this peak-matching method of "tuning" the time scale.

According to the Fisher *et al.* chronology for A84, the top half of the core covers the years A.D. 1984 (when the core was drilled) back to about A.D. 1080.[10] That is, the top half of the core covers just 900 years. The lowest large ECM peak in the core, at about 59 meters above bedrock, they date to A.D. 934. This much of the chronology is probably reasonably accurate. But the section of core below this final large ECM peak, back to the end of the last glacial period at roughly 12.3 meters above bedrock, spans roughly 12,500 years according to the Fisher *et*

[10]David A. Fisher, Roy M. Koerner, and Niels Reeh, "Holocene Climatic Records from Agassiz Ice Cap, Ellesmere Island, NWT, Canada," *The Holocene* 5.1 (1995): 21, Table 2.

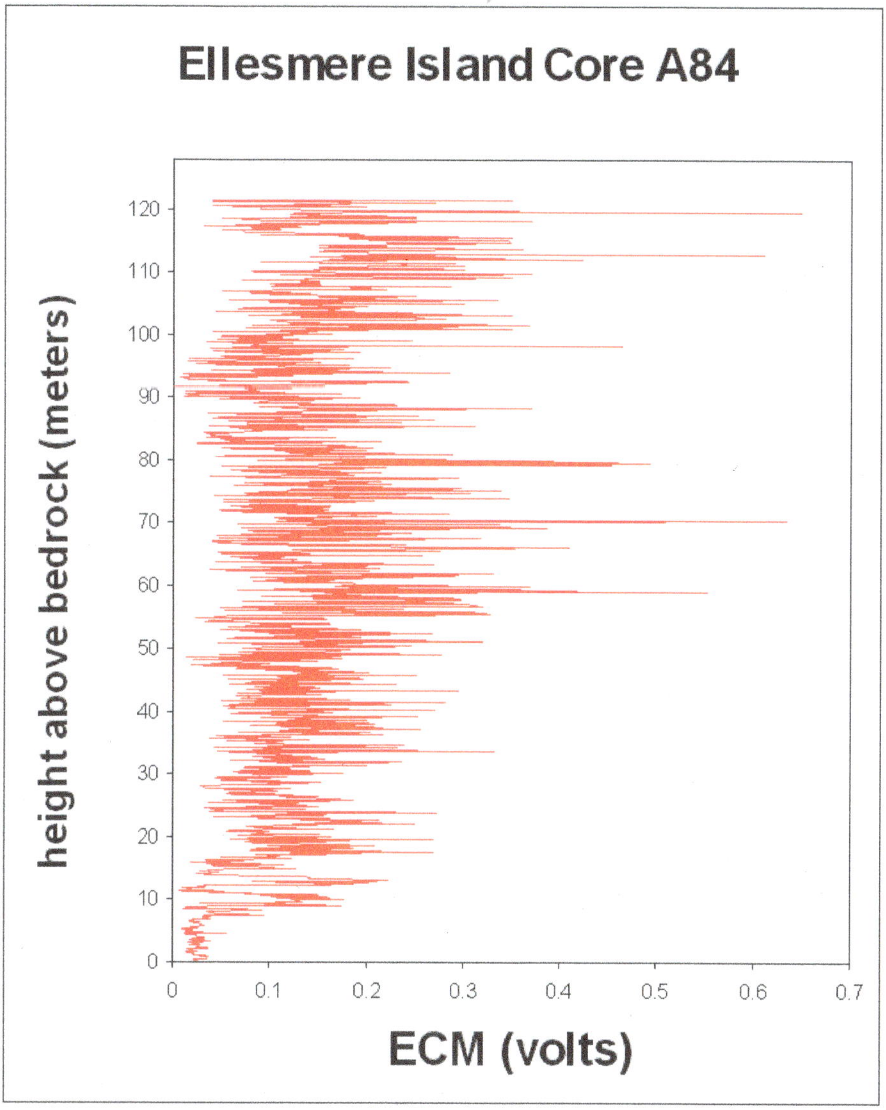

Figure 20.3: Electrical conductivity measurement data for the A84 ice core from Ellesmere Island, Canada, from bedrock to nearly the top of the core at 127 meters.
(Data were obtained via http://www.ncdc.noaa.gov/paleo/icecore/polar/ Agassiz/ecm.html. The graph was constructed using all of the data from the file a84ecm_5cm.txt.)

al. chronology, and this 12,500 years is devoid of ECM peaks which are clearly above background. The best candidate for volcanic peak matching in the lower portion of the core is the peak at 33.5 meters above bedrock, but Fisher *et al.* are unable to match any volcano to this peak.

It appears that almost any chronology of the deeper portion of A84 could be "tuned" to match peaks in these data to peaks in a table of dated volcanic eruptions. Subjective leeway seems intrinsically large with this method of dating this ancient ice.

Radiocarbon Dates

To help guard against subjective bias, it is possible to use radiocarbon dating on ice cores. To use radiocarbon to provide the entire chronology for an ice sheet would be much too labor intensive and expensive, and not as precise as layer counting in any event. But by radiocarbon dating a few samples at various heights along the core an independent check of a time scale computed by other means can be obtained. This was done with the Devon Island ice cap chronology, for example.[11] But publications describing construction of the chronologies of the Agassiz cores make no mention that I have been able to find of any radiocarbon dates.[12]

Two δ-Step Mechanisms

There appears to be no good reason for confidence in the published chronology of A84. Its validity has really been rested in its rough congruence with the "known" end-of-glacial-period date of the δ step (Figure 20.2, page 149), yet it is this "known" which is called most seriously into question by the historical reality of the Flood.

[11]W. S. B. Paterson, R. M. Koerner, D. Fisher, S. J. Johnsen, H. B. Clausen, W. Dansgaard, P. Bucher, and H. Oeschger, "An Oxygen-isotope Climatic Record from the Devon Island ice cap, Arctic Canada," *Nature* 266 (7 April 1977): 508–511.

[12]D. A. Fisher, R. M. Koerner, W. S. B. Paterson, W. Dansgaard, N. Gundestrup, and N. Reeh, "Effect of Wind Scouring on Climatic Records from Ice-core Oxygen-isotope Profiles," *Nature* 301 (20 January 1983): 205–209. D. A. Fisher and R. M. Koerner, "The Effects of Wind on $\delta(^{18}O)$ and Accumulation give an Inferred Record of Seasonal δ Amplitude from the Agassiz Ice Cap, Ellesmere Island, Canada," *Annals of Glaciology* 10 (1988): 34–37. David A. Fisher, Roy M. Koerner, and Niels Reeh, "Holocene Climatic Records from Agassiz Ice Cap, Ellesmere Island, NWT, Canada," *The Holocene* 5.1 (1995): 19–24.

While the assumption of an end-of-glacial-period date for the sudden δ step in the A84 core is pretty much unavoidable within the current conventional scientific mind-set, it is called into question as soon as it is realized that Noah's Flood was a real historical event which submerged these ice caps for several months 3520 B.C. Given the Flood, a sudden change in δO^{18} may be due to the end of the last glacial period, or it may be due to meltback of the pre-Flood ice cap by the water of the Flood.

If the Flood melted away all of the pre-Flood Holocene ice and some of the glacial period ice at the A84 borehole location, then whatever δ step may have been present due to the end of the glacial period would have been erased. New (post-Flood) mid-Holocene ice, with high δO^{18}, would then have been deposited directly on top of old glacial period ice having low δO^{18}. Thus, discontinuous deposition at the meltback horizon would be the cause of the sudden δ step presently observed in the A84 core, not retreat of the ice sheets at the end of the glacial period.

Notice that core A84 is only 127 meters high today. It was previously estimated that the Flood caused roughly 140 meters of meltback at Devon Island. It is thus entirely possible that the Agassiz ice cap might have lost its entire Holocene ice accumulation and a significant part of its glacial period accumulation to meltback by the Flood at the location of borehole A84.

There are, therefore, two mechanisms for producing large δ steps once the Flood is allowed its rightful place in history: (1) end-of-glacial-period ice sheet retreat, and (2) meltback occasioned by the Flood.

A Simple Test

Which of these two mechanisms is more likely to have caused the δ step observed in core A84 can be tested in a fairly easy way. The two mechanisms differ substantially in regard to the rapidity of their action.

The Flood mechanism places mid-Holocene snow directly on top of old glacial period ice. Even allowing for some diffusion of isotopes over the past 5500 years, the transition from lower to higher δO^{18} values should be visibly sudden in this case.

In contrast to this, retreat of the ice sheets at the close of the last glacial period could not have happened in a single year. The transition from glacial period to Holocene ice sheet thicknesses would necessarily have taken many centuries to accomplish. Thus the transition from lower to higher δO^{18} values should be visibly gradual in this case.

Figure 20.4 shows a comparison of the bottom 15 meters in both the Devon Island D73 core and the Ellesmere Island A84 core. Notice how very different the shapes of the two curves are. The Devon Island curve (D73) makes a sudden jump to the left and then slowly returns to the right, with some smaller excursions along the way. The Ellesmere Island curve (A84) does the opposite. It slowly moves to the left, with various smaller excursions along the way, then suddenly jumps to the right. Is it reasonable to suppose that both of these curves display a record of the same phenomenon?

The transition from lower to higher δO^{18} values is visibly gradual for Devon Island, as expected. This is a record of gradual thinning of the ice sheet at the transition from glacial period to Holocene.

In contrast, the transition from lower to higher δO^{18} values is visibly sharp for Ellesmere Island. This is not a record of gradual thinning. There is no record of gradual thinning in this ice core. That part of the record is missing. What remains is a record of sudden discontinuity—deposition of snow having higher δO^{18} directly on top of ice having much lower δO^{18}.

The evidence is simply contrary to the claim that the δ step in A84 is due to retreat of the ice sheets at the end of the glacial period.

A Final Confirmation

This conclusion is strengthened by comparing the δ step in the Devon Island core with the δ step in a core from Greenland.

The Greenland ice sheet is very thick (several kilometers), so the end-of-glacial-period transition in question would have been safe from the Flood, protected deep inside the core from meltback of one or two hundred meters either topside or bottom-side. The δ step for such a core will surely be due to the end-of-glacial-period mechanism, just as it is in the Devon Island core. Thus both of these cores should show the same basic shape for the δ step, recording the changes which were taking place in the Arctic region at the end of the last glacial period.

Figure 20.5 (page 160) shows that the shape of the δ step can be matched in considerable detail between Devon Island and the Dye 3 core from Greenland. (See Figure 20.1, page 147, for the location of Dye3.)

The lowest 2.6 meters of the D73 core do not match, implying a

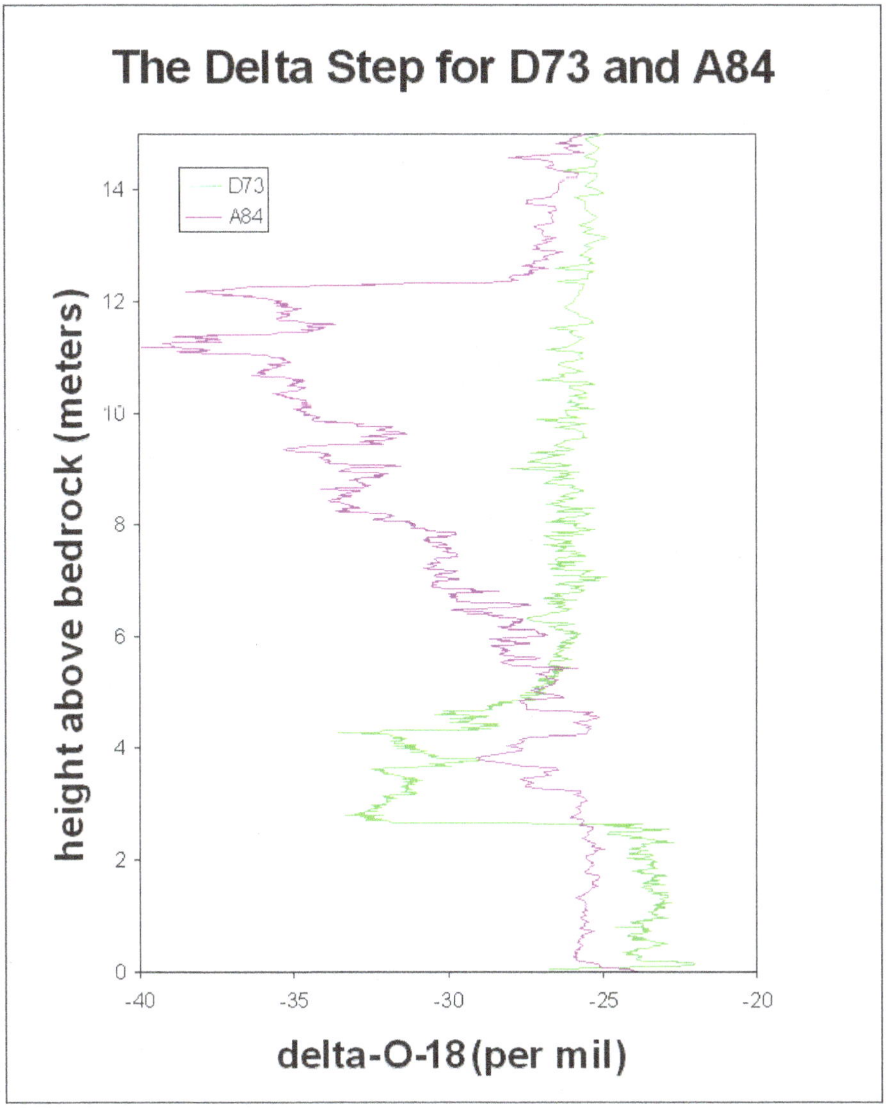

Figure 20.4: Comparison of the δ step for the Devon Island D73 core and the Ellesmere Island A84 core.
(Data for D73 were obtained via http://www.ncdc.noaa.gov/paleo/icecore/polar/devon/isotope.html using data from the files D73delws_200yr.txt and d73delde_50yr.txt. Data for A84 were obtained via http://www.ncdc.noaa.gov/paleo/icecore/polar/Agassiz/raw.html from the files mer84-25_a84.txt and mer84-ws_a84.txt.)

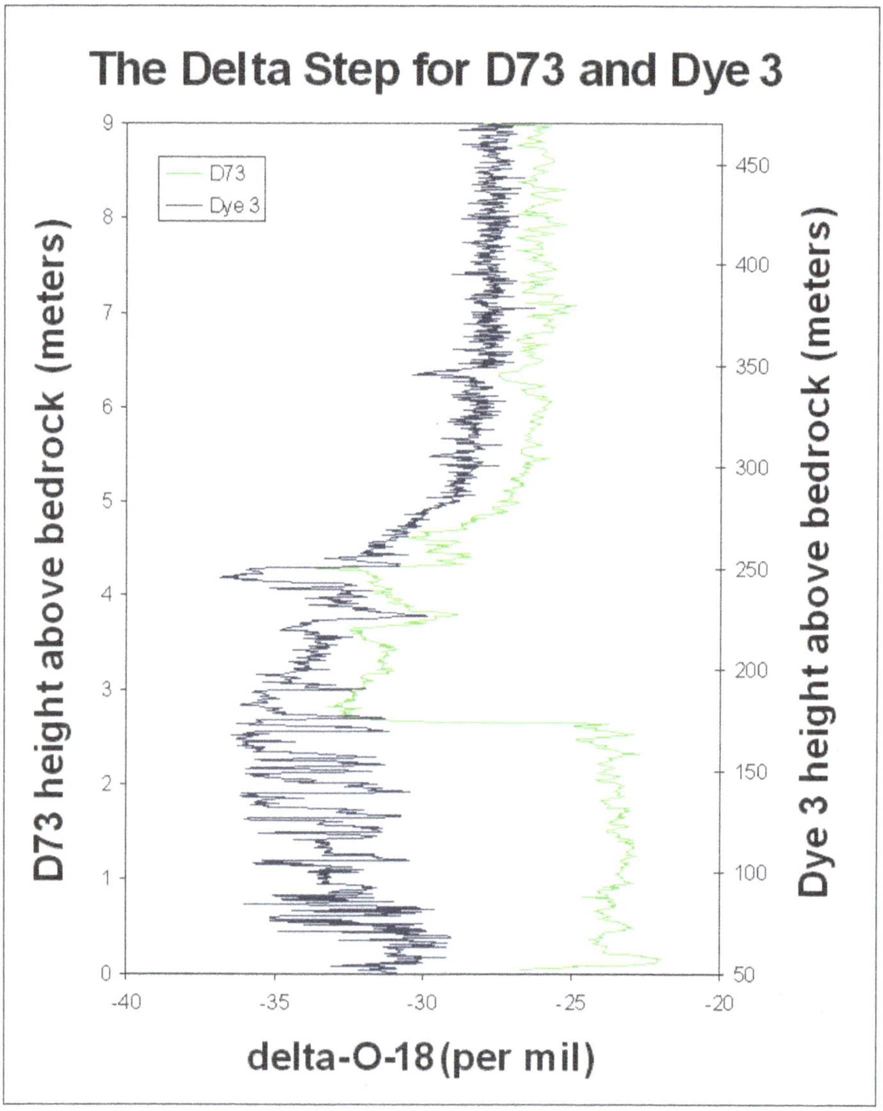

Figure 20.5: Comparison of the δ step for the Devon Island D73 core and the Greenland Dye 3 core.
(Data for D73 were obtained via http://www.ncdc.noaa.gov/paleo/icecore/polar/devon/isotope.html using data from the files d73delws_200yr.txt and d73delde_50yr.txt. Data for Dye 3 were obtained via http://www.ncdc.noaa.gov/paleo/icecore/Greenland/gisp/dye3/dye3_data.html from the file dye3.txt.)

disturbance of some sort of these lowest 2.6 meters of the D73 core. But above this disturbance there is good agreement.

The region of agreement appears to cover the entire transition from glacial period to Holocene, thus establishing the general shape of the δ step which should be expected from this transition in Arctic ice sheets. That this shape is reproducible between widely separated cores is unambiguously demonstrated. If the A84 δ step were due to the end-of-glacial-period transition, it would have this shape. Since it clearly does not have this shape, it is not due to the end-of-glacial-period transition.

Summary

There is general agreement by all parties that the ice cap appears to have suffered topside meltback. The only issue is the cause of this meltback.

The original researchers dated the meltback to the early Holocene. If this date is correct then the meltback is not due to Noah's Flood because Noah's Flood happened mid-Holocene.

The original researchers dated the loss of ice to the early Holocene because they assumed that the δ step present in the cores taken from the Agassiz ice cap on Ellesmere Island must be due to the end of the last glacial period. They assigned this δ step a date in excess of ten thousand years ago based on this assumption. They proceeded to work out a chronology of the ice cap in basic harmony with this assumption.

But the shape of the δ step contradicts the assumption that it is due to the end of the last glacial period. It is now clear that the record of the transition from glacial period to Holocene is entirely missing from the Agassiz ice cap. The δ step actually shows deposition of Holocene snow directly on top of ancient glacial period ice. Thus the assumption of coincidence of the δ step with the end of the last glacial period is not supported. And with the loss of this assumption—the loss of this "known" point—the entire chronology which has been built upon it collapses. And when this chronology collapses, the claim of early Holocene meltback which rests upon it also collapses.

Conclusion

The objection that Noah's Flood is missing at Ellesmere Island is not substantiated. It appears that Noah's Flood covered this entire region 5500 years ago.

CHAPTER TWENTY-ONE

NOAH'S FLOOD AT ELK LAKE, MINNESOTA

Elk Lake in Itasca State Park in Minnesota (Figure 21.1) is a special lake for the scientific study of the past. In contrast to most lakes, it has well-preserved laminated sediments. This results from the fact that Elk

Figure 21.1: The red X marks the location of Elk Lake, Minnesota. (Map adapted from *Reader's Digest Great World Atlas*, 1963.)

Lake is quite deep relative to its surface area. Its surface area (1 square kilometer or 250 acres) is average for lakes in Minnesota, but its depth is unusual. It is a very deep lake with a 30-meter (97-foot) depression in its southeastern end today.

In the past, the lake was considerably deeper. Laminated lake sediments have raised the bottom of the lake in the depression from an original 50 meters (160 feet) to the present 30 meters.

The unusual depth to surface area ratio of the lake makes it difficult for natural processes to supply the deepest parts of the lake with oxygen. As a result, macroscopic organisms such as fish, worms, and clams, which need oxygen to live, and which constantly mix the sediments at the bottom of most lakes, are unable to do so at the bottom of Elk Lake.

A lake's sediments are rich with certain kinds of information about its past. For example, pollen spores recovered from the sediments reveal the types of vegetation which surrounded the lake in the past. Past windiness in the vicinity of the lake can be deduced from the amount and character of dust found within the lake's sediments. Other measurable parameters having distinct contributions to make to the story of the lake's past include, for example: types of diatoms which grew within the lake, elements and stable isotopes of sedimentary components, bulk magnetic properties, and fossil pigments.

Elk Lake's well-preserved, laminated sediments greatly enhance its story of the past because they provide it with a simple chronological framework. At Elk Lake, it is possible to deduce with some assurance not only *what* happened, but also *when* it happened.

The Flood has been found in the Middle East, in Ireland, and in the Arctic (Figure 21.2). It is by now clear that the Flood was not in the same category as the floods that are familiar to us in modern times. It was very much larger in extent and very much deeper than any modern flood. Can this monstrous Flood be tracked yet farther around the globe? Did it extend to Elk Lake, Minnesota?

Once again, the answer is yes.

Predicting the Flood's Impact at Elk Lake

What should evidence of the Flood be expected to look like today, if the Flood extended to Elk Lake?

It has already been seen, in the Middle East and in Ireland, that the

Figure 21.2: View of Earth from above the North Pole, showing the location of Elk Lake relative to other locations, previously discussed, in which evidence of the Flood has been found. (M = Mesopotamia, P = Palestine, I = Ireland, D = Devon Island, E = Ellesmere Island, L = Elk Lake; figure adapted from *Reader's Digest Great World Atlas*, 1963.)

Flood did not tear up and redeposit the surface of the ground. Because Elk Lake is very deep, it seems reasonable to expect that pre-Flood sedimentary layers from the deepest parts of the lake would have survived the Flood undisturbed.

In addition, floods are ideal agents for making sediments. Elk Lake is a deep basin, ideal for catching and preserving sediments. Thus, an unusually thick layer of sediment due to the Flood is perhaps the most obvious expectation.

Beyond this, prediction becomes increasingly precarious. How thick such a layer should be, and what it should look like in detail, are matters impossible to guess at this stage.

The difficulty is that no fixed scientific model of the Flood exists at this stage of the investigation to help elucidate what the Flood's action at Elk Lake might have been like. Indeed, at this stage, the investigation is focused on just the opposite activity. Data on the extent of the Flood are presently being gathered to enable eventual formulation of a reasonably accurate scientific model of the Flood.

The chronology is the only real known at this stage. All that can be said with confidence is that, if the Flood extended to Elk Lake, then the Elk Lake sedimentary data may possibly show some kind of anomaly within dating uncertainties of 3520±21 B.C.

Do the Elk Lake data show any anomaly at this date?

Yes, they most certainly do.

Elk Lake's Sediments

Edward B. Nuhfer *et al.* studied the processes which produce laminated sediments in Elk Lake today.[1] They collected sediment as it settled from the overlying water column in Elk Lake. They used specially designed sediment traps equipped with a time marking device so they could record when different types of sediment settled out. The traps were placed in Elk Lake in the late seventies and early eighties.

[1]Edward B. Nuhfer, Roger Y. Anderson, J. Platt Bradbury, and Walter E. Dean, "Modern Sedimentation in Elk Lake, Clearwater County, Minnesota," *Elk Lake, Minnesota: Evidence for Rapid Climate Change in the North-Central United States*, ed. J. Platt Bradbury and Walter E. Dean (Boulder: The Geological Society of America, Inc., 1993) 75–96.

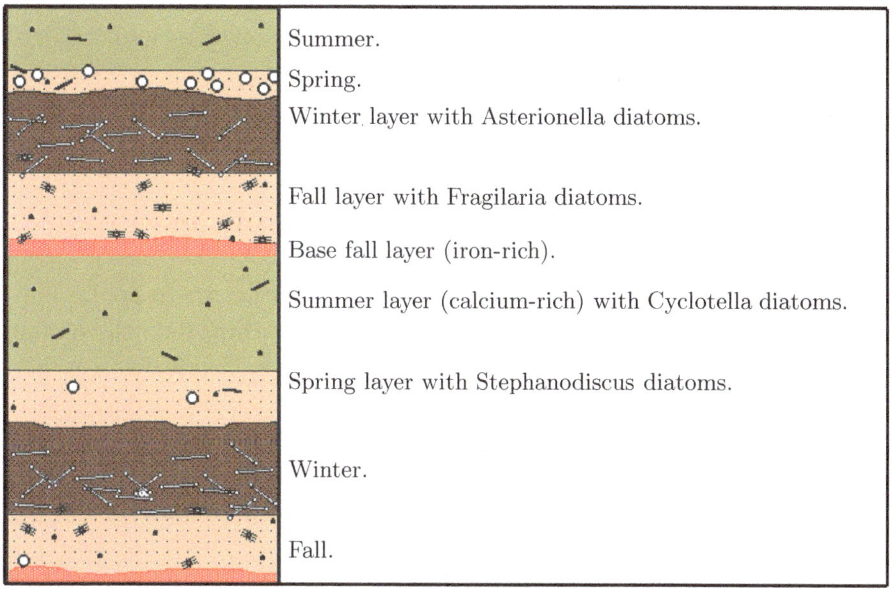

Figure 21.3: Idealized portion of sediment column from Elk Lake showing two years of accumulation. [Adapted from Figure 11 of E. B. Nuhfer *et al.* See footnote 1 for full reference.]

Nuhfer *et al.* found that "modern laminations in Elk Lake are created by distinct seasonal processes."[2] In other words, the sediments which are accumulating at the bottom of Elk Lake today are laminated because of the yearly cycle of seasons. This comes about as follows.

In the winter, the lake freezes over. The ice cover typically lasts for about five months. During this time, the water in the lake does not circulate, and organic detritus mixed with precipitates of iron and manganese settles to the bottom of the lake producing a relatively thick brown layer of sediment.

In the spring, following the melting of the ice, wind-driven circulation of the lake takes place until the lake becomes thermally stratified. The period of circulation is variable, depending on how warm and windy the spring is. During this brief period, sediment is resuspended from shallower portions of the lake, producing a layer of redeposited sediment in

[2]Edward B. Nuhfer, Roger Y. Anderson, J. Platt Bradbury, and Walter E. Dean, "Modern Sedimentation in Elk Lake, Clearwater County, Minnesota," *Elk Lake, Minnesota: Evidence for Rapid Climate Change in the North-Central United States*, ed. J. Platt Bradbury and Walter E. Dean (Boulder: The Geological Society of America, Inc., 1993), 93.

the deep portions of the lake. The thickness of this layer varies considerably from year to year.

Summer stratification brings stagnation once again, which persists for another four or five months. During this period, the warmer temperatures in the upper water layers of the lake cause calcium carbonate to precipitate, producing another relatively thick, distinct, characteristically light-colored sediment layer.

In the fall, the warm surface layers of the lake begin to cool. This causes iron compounds to precipitate, producing a thin, reddish-colored sediment layer. Fall winds mix the lake once again, producing another redeposited layer similar in its characteristics to the spring layer.

Thus, at the present time, five sediment layers are typically produced each year in a repeating cycle (Figure 21.3, page 167).

The individual seasonal sediment layers also contain a biological component which exhibits a well-defined annual cyclicity. Specifically, diatoms (i.e., minute unicellular algae with silica shells, called frustules) live in the lake. While these are present year round, the spring and fall circulations provide them most abundantly with the essential elements they require for growth and reproduction. Thus, the sediment layers produced at these times of the year are especially rich in diatom frustules. In fact, in some years diatom blooms can be sufficiently extensive to produce spring or fall sediment layers which are made up almost exclusively of diatom frustules.

Each individual type of diatom has its own unique preference for temperature and other environmental factors. Thus, spring and fall diatom blooms tend to be dominated by different diatom species.

At the present time, about 2 millimeters (slightly less than one tenth of an inch) of sediment accumulates each year. The individual seasonal layers are, therefore, only about one half millimeter thick. Nonetheless, these are easily resolved under low-power magnification, and their entrained diatoms can be easily identified using higher-power magnification.

The counting of bundles of seasonal sediment layers, corresponding to a single year, is the basis for the chronology at Elk Lake. Such "direct counting" methods of chronology building are very labor intensive. Imagine the work involved in collecting cores from the bottom of the lake (done with piston corers through holes in the ice during the winter) and then processing these cores (freezing, halving, surfacing, photographing) to enable counting the 2-millimeter-thick layers through 20 meters of bottom sediment!

But the work required has been done in the present case, and the result has proven to be well worth the effort, yielding a chronology for the cores and hence for the sedimentary history of the lake.

In fact, about 10,000 bundles of seasonal layers have been counted by the original researchers at Elk Lake from the bottom to the top of the 20 meters of sediments which have accumulated at the bottom of the lake. Said simply, Elk Lake appears to have recorded approximately 10,000 years of history in its sediments.

The Elk Lake Data

Figure 21.4 (page 170) shows measured thicknesses of what the original researchers believed to be annual sedimentary layers at Elk Lake over the entire record.[3] The original researchers have called these layers "varves." I am deliberately avoiding use of the word "varves" because it implies annual cycles of deposition, and it is not at all clear at this stage that all of the counted layers are annual. As usual, there is a component of subjectivity involved in determining what constitutes an annual layer, in addition to difficulties of a technical sort, so I am opting to use the time-neutral term "layer" instead of "varve," asking the reader to bear in mind that each layer is comprised of multiple laminations as discussed above.

The immediate goal is to be able to match the sedimentary layer data of Figure 21.4 to the biblical chronology date of the Flood, 3520±21 B.C. It is most unlikely that the layer-counting techniques used to construct the chronology in Figure 21.4 result in a dating precision comparable to the ±21 years afforded the date of the Flood by biblical chronology. Fortunately, the original researchers have carried out a number of radiocarbon measurements on organic material found within the sediments of the cores, and these measurements can be used to correct for inadvertent subjective bias and to improve overall accuracy of the chronology. I will address this shortly, but first it is important to gain a general familiarity with the sedimentary data as they have been presented by the original researchers.

[3]For greatest visual clarity, only the thickest and thinnest layers are plotted from each consecutive group of twelve layers. (Data supplied by Bruce Bauer, National Geophysical Data Center, National Oceanic and Atmospheric Administration.)

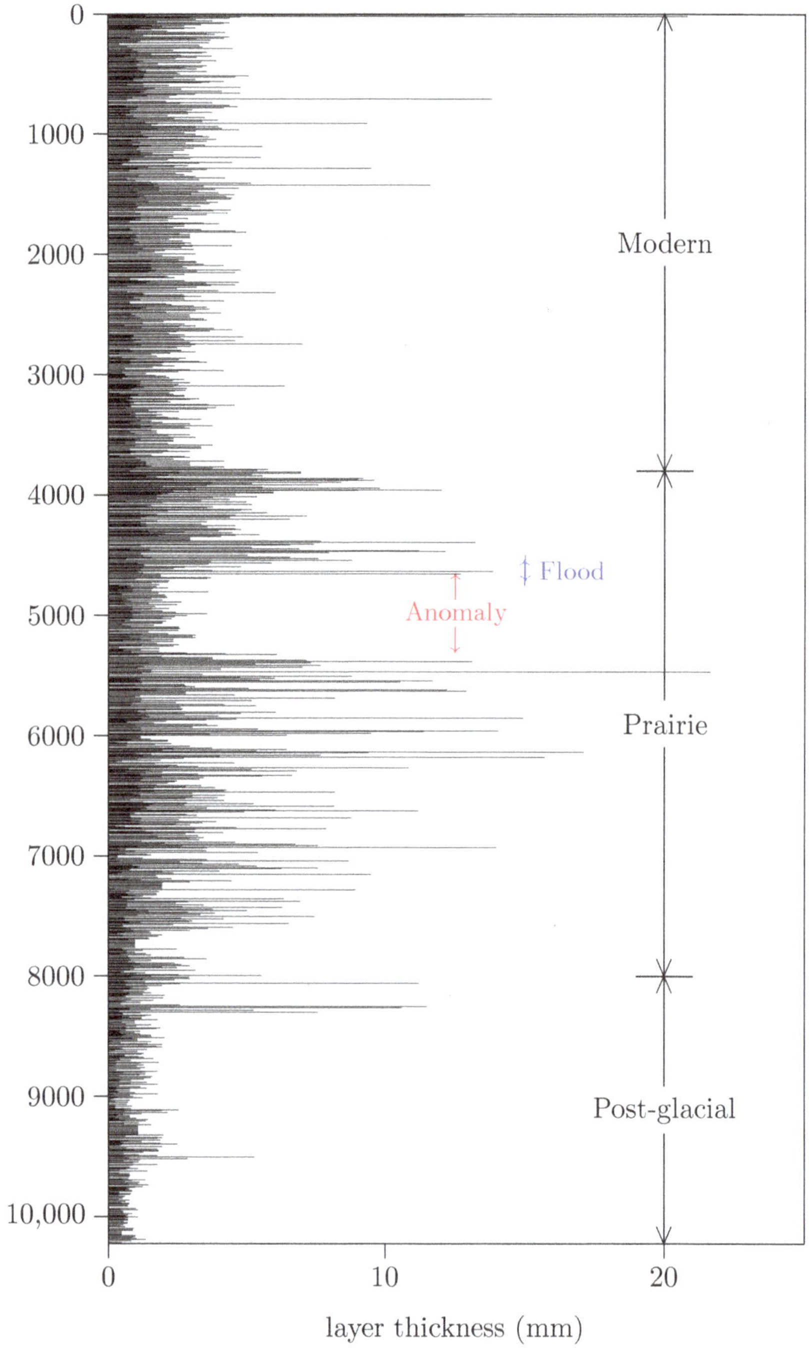

Figure 21.4: Elk Lake sedimentary layer thickness.

Those who have studied the wealth of data preserved in the sediments of Elk Lake most closely divide the history of the lake into the three stages shown in Figure 21.4. Roger Y. Anderson of the department of Earth and Planetary Sciences of the University of New Mexico summarizes these three stages as follows (square brackets in this and subsequent quotations indicate amplification by me):[4]

> About 14 ka [i.e., 14,000 years ago], withdrawal of the [glacial] ice sheet stranded [and buried] a large block of ice that had melted by about 3000 yr later. The ice block formed a deep lake basin in a terrain of unweathered till, clothed by coniferous forest [as revealed by pollen from the sedimentary layers produced in the lake at that time]. This early [post-glacial] lake and its environs was dominated by cold, anticyclonic [i.e., high pressure system] winds from the [glacial] ice margin to the northeast...
>
> About 8.0 ka, after further decay of the ice sheet, cold anticyclonic winds were replaced by incursions of relatively dry Pacific air that reached progressively farther into the continental interior. In response to these incursions, the prairie shifted northeastward and there ensued a 4000 yr period of drought in north-central Minnesota [where Elk Lake is located]. The more saline mid-Holocene [prairie] lake was surrounded by scattered stands of oak, sparse grass, *Artemisia* [i.e., sagebrush], and some open, bare ground. The varves [i.e., annual layers of sediment] that accumulated in Elk Lake more than doubled in thickness as a result of the influx of eolian [i.e., wind-borne] clay and silt. The loess [i.e., wind-produced deposit] was suspended from the region to the west and was carried to Elk Lake by dry westerly winds...
>
> Within a few centuries, at about 3.8 ka, the tropical airstream [from the Gulf of Mexico] moved northward, bringing additional moisture, and a new balance was struck between Arctic,

[4]Roger Y. Anderson, "The varve chronometer in Elk Lake: Record of climatic variability and evidence for solar-geomagnetic-^{14}C-climate connection," *Elk Lake, Minnesota: Evidence for Rapid Climate Change in the North-Central United States*, ed. J. Platt Bradbury and Walter E. Dean (Boulder: The Geological Society of America, Inc., 1993), 45–46.

> Pacific, and Tropical airstreams. The new expression of moisture and seasonality brought pine, hardwoods, and forest soils to the Elk Lake drainage.

This brief history reveals a long-term progressive change in the climate of Elk Lake from the end of the glaciation which created the lake to the present time. This long-term change seems to be adequately explained by two factors: the retreat of the ice sheet at the end of the last glaciation, and changes in insolation (i.e., total radiation received from the sun) due to slow, long-term changes in earth's tilt and orbital parameters. Computer simulations which take these parameters into consideration support this conclusion, as the following observations from one such simulation show:[5]

> When the ice sheet was large, generally cool conditions should have prevailed, consistent with the generally cool conditions inferred from the pollen data from Elk Lake for the interval 11,600–6000 varve yr. With the replacement of the glacial anticyclonic wind regime by stronger westerlies during the interval between 9000 and 6000 varve yr, precipitation should have decreased to its lowest levels during the Holocene, again consistent with the Elk Lake evidence. Finally, during the past 6000 yr, modern conditions should have developed as the seasonal distribution of insolation gradually approached present-day levels.

Most of the details of the sedimentary record are also readily explained within this overall climatic framework. For example, sediment layers are observed to be highly variable and often thick during the prairie period (Figure 21.4, page 170). The pollen data (e.g., of sagebrush) clearly indicate that this was a dry period at Elk Lake. Several factors can be identified which conspire to produce thick and variable annual layers when the climate is dry. First, the lake becomes surrounded by sparsely vegetated, open prairie. Much soil is exposed to the elements. The ground is easily dried by the sun, and prevailing winds, not blocked near the ground by forest, are able to pick up clay and silt from the

[5]Patrick J. Bartlein and Cathy Whitlock, "Paleoclimatic interpretation of the Elk Lake pollen record," *Elk Lake, Minnesota: Evidence for Rapid Climate Change in the North-Central United States*, ed. J. Platt Bradbury and Walter E. Dean (Boulder: The Geological Society of America, Inc., 1993), 288–289.

ground and deposit them in the lake. This wind blown detritus adds directly to the annual layer thickness of lake sediments, of course, but it also adds indirectly to the layer thickness by fertilizing the lake and increasing its own biologically produced sediment load. In addition to these factors, the level of the lake reduces when the climate is dry. This enhances the ability of waves produced by winds (again not blocked by forest) to resuspend sediment from the shallow margins of the lake and redeposit it in the deep interior of the lake (from which the sediment cores were taken).

Conditions are more moist today than they were back in the prairie period. As a result, the lake is surrounded by forest, which limits production of wind-borne detritus and inhibits formation of thick annual layers. Thus the fact that annual layers are relatively thin and stable for the modern lake stage is also easily understood, and the transition from thick, variable layers to thin, stable layers which begins shortly after (i.e., higher up than) layer 4000 and takes several centuries to complete, is also easily explained.

The Anomaly

What is not explained—what is anomalous—is the sudden transition into a thin, stable layer production mode at about layer 5300, and the equally sudden transition back out again at about layer 4700 (Figure 21.4, page 170). What combination of geophysical and climatic factors could have produced these unexpected 600 layers? This question surfaces repeatedly within the 336-page Geological Society of America report which summarizes the extensive research which has been carried out on the lake. I think that at least some of the original researchers may have convinced themselves that they have adequately understood this anomaly. But no truly satisfactory answer to this question is ever given within the report.[6]

Once it is understood that thick, variable layers result from dry conditions, it is natural to suppose that the 600 thin uniform layers must represent a period of temporarily increased moisture. This, in fact, is what the original researchers proffer as the explanation of this anomaly:[7]

[6]J. Platt Bradbury and Walter E. Dean, ed., *Elk Lake, Minnesota: Evidence for Rapid Climate Change in the North-Central United States* (Boulder: The Geological Society of America, Inc., 1993).

[7]J. Platt Bradbury, Walter E. Dean, and Roger Y. Anderson, "Holocene climatic and limnologic history of the north-central United States as recorded in the varved

> The long-term changes in soil moisture that [hypothetically] at first increased and then lowered the threshold, shutting off and then turning on dust suspension, probably were more gradual than indicated by the event [i.e., the anomaly] itself, occurring in an interval that was longer than 600 yr. This suspected longer interval must have been accompanied by a significant increase and later decrease in moisture in order to account for the event.

But this "significant increase in moisture" theory doesn't really work. The difficulty is that the data from the cores during this 600-layer interval do not seem to paint a coherent picture.

For example, the average thickness of the sediment layers during this interval and their low variability are characteristics which are most closely matched by the annual layers of the modern lake stage. This implies a climate more similar to that of today than to that of the prairie stage in which these 600 layers are embedded, making increased moisture seem reasonable. But the pollen data show that there was no change in the vegetation surrounding the lake during the interval in which these layers were deposited. As far as the vegetation was concerned, the lake's environment was still prairie for the entire time:

> At Elk Lake the prairie period has been divided into three climate phases on the basis of diatom, sedimentologic, and geochemical data: an early xeric [i.e., low moisture] phase between 8500 and 5400 varve yr, a somewhat wetter phase from 5400 to 4800 varve yr [the anomaly], and a dry phase between 4800 and 4000 varve yr. The pollen record does not show this subdivision clearly...[8]

> At this time the lake entered a brief phase [the anomaly] (ca. 600 yr) that appears to have been a precursor to lake conditions characteristic of the latest stage in Elk Lake's development beginning at 3.8 ka, but differs from the more permanent

sediments of Elk Lake, Minnesota: A synthesis," *Elk Lake, Minnesota: Evidence for Rapid Climate Change in the North-Central United States*, ed. J. Platt Bradbury and Walter E. Dean (Boulder: The Geological Society of America, Inc., 1993), 319.

[8]Cathy Whitlock, Patrick J. Bartlein, and William A. Watts, "Vegetation history of Elk Lake," *Elk Lake, Minnesota: Evidence for Rapid Climate Change in the North-Central United States*, ed. J. Platt Bradbury and Walter E. Dean (Boulder: The Geological Society of America, Inc., 1993), 258.

> change because it was not accompanied by the same changes in vegetation.[9]

To further add to the complexity of the situation, cysts from chrysophycean algae show a return to conditions which existed in the *post-glacial* lake during this interval:[10]

> Our cyst-assemblage data indicate that limnological [i.e., lake] conditions during the mid-Holocene prairie period shifted for a period of about 500 yr between 5.3 and 4.8 ky. Dominant cysts indicate a striking return to early postglacial conditions...

How is it possible for a single lake to go on showing characteristics of three very different climatic stages all at the same time for 600 years? This period is truly anomalous, and it is the only real anomaly in the entire dataset.

The Anomaly and the Flood

Might this anomaly be due to the Flood?

Getting the answer to this question obviously has everything to do with chronology. A causal relationship can only be postulated if there exists a temporal coincidence between the Flood and the anomaly. The single most important question at this point, therefore, is: "Does the Flood coincide with the Elk Lake anomaly?"

Chronology of the Elk Lake Sediments

Biblical chronology places the date of the Flood at 3520±21 B.C. The topmost annual sedimentary layer which has been preserved in core sam-

---

[9] J. Platt Bradbury, Walter E. Dean, and Roger Y. Anderson, "Holocene climatic and limnologic history of the north-central United States as recorded in the varved sediments of Elk Lake, Minnesota: A synthesis," *Elk Lake, Minnesota: Evidence for Rapid Climate Change in the North-Central United States*, ed. J. Platt Bradbury and Walter E. Dean (Boulder: The Geological Society of America, Inc., 1993), 318.

[10] Barbara A. Zeeb and John P. Smol, "Postglacial chrysophycean cyst record from Elk Lake, Minnesota," *Elk Lake, Minnesota: Evidence for Rapid Climate Change in the North-Central United States*, ed. J. Platt Bradbury and Walter E. Dean (Boulder: The Geological Society of America, Inc., 1993), 247.

ples from Elk Lake (i.e., annual layer 1) corresponds to A.D. 1927.[11] Thus, if there were no counting error or dating uncertainty of any sort, 3520 B.C. would correspond to (1927+3520=) layer 5447. In fact, however, there are counting and dating uncertainties which prohibit the identification of the year of the Flood with any single sedimentary layer. These uncertainties allow the Flood to be located anywhere within a specified range of layers.

There is, first of all, an uncertainty in the biblical date of the Flood of ±21 years (3σ). This is quite tiny, however, relative to the experimental counting uncertainty of annual layers at Elk Lake. Layer counting uncertainty is due mainly to technical aspects of preservation and recovery of the annual sedimentary layers during coring and processing by the science laboratory. Donald R. Sprowl has measured this uncertainty by comparison of annual layer counts in independent cores covering the same time interval. He has found that it amounts to about ±500 layers (2σ) at the 5447th annual layer.[12] Therefore, if this is the most significant source of error, then the year of the Flood should be expected to fall with near certainty within ±750 layers (3σ) of layer 5447 (i.e., somewhere between layer 4700 and layer 6200).

However, there is another potential source of error which significantly affects the effort to compare the biblical chronology date of the Flood with the secular chronology at Elk Lake. Sprowl explains:[13]

> The above analysis assumes that the counting errors are normally distributed with zero mean, but this is only true if the probability of counting too many varves [i.e., annual sedimentary layers] is the same as that of counting too few. Too many varves can be counted when subannual sets of lamina-

---

[11] Roger Y. Anderson, J. Platt Bradbury, Walter E. Dean and Minze Stuiver, "Chronology of Elk Lake sediments: Coring, sampling, and time-series construction," *Elk Lake, Minnesota: Evidence for Rapid Climate Change in the North-Central United States*, ed. J. Platt Bradbury and Walter E. Dean (Boulder: The Geological Society of America, Inc., 1993), 40.

[12] Donald R. Sprowl, "On the precision of the Elk Lake varve chronology," *Elk Lake, Minnesota: Evidence for Rapid Climate Change in the North-Central United States*, ed. J. Platt Bradbury and Walter E. Dean (Boulder: The Geological Society of America, Inc., 1993), 72.

[13] Donald R. Sprowl, "On the precision of the Elk Lake varve chronology," *Elk Lake, Minnesota: Evidence for Rapid Climate Change in the North-Central United States*, ed. J. Platt Bradbury and Walter E. Dean (Boulder: The Geological Society of America, Inc., 1993), 74

> tions appear to be annual. However, varves can be obscured or left undistinguished sedimentologically or by the cleaning or polishing technique used. It is my (subjective) judgment that counting too few varves is more likely than counting too many, and I expect the errors to be biased on the low side. ... Because of this expected negative bias in the varve counting process, the highest count from a given interval was used as the estimator of the actual number of varves present. Presumably, this still underestimates the actual number of varves present.

This potential loss of annual layers must also be taken into consideration if the comparison of biblical and Elk Lake chronologies is to be meaningful. Notice that if ten percent of the layers have been missed (which doesn't seem at all impossible, judged on the basis of the measured counting precision) the layer number will be 550 years short of true calendar years by the time of the Flood (which was roughly 5500 years ago). This is a substantial offset, which cannot be ignored.

This problem cannot be solved by counting annual layers in duplicate cores because the same problem of unidentified (missing) annual layers will pertain to both cores. An independent chronometer is needed, one which can be applied to the Elk Lake data to help determine just how many annual layers may have been missed in the counting process. Fortunately, radiocarbon supplies what is needed in this instance.

Sixteen radiocarbon measurements were made on organic carbon from the Elk Lake cores.[14] These reveal (Appendix C, page 325) that 17.6% of annual layers have been missed in the layer counting process for the Modern lake stage. When these missing layers are corrected for, it is found that the Flood is expected to be seen, if present in the Elk Lake data, at layer number 4630 ± 90 (3σ).

This is the range shown by back-to-back arrows labeled "Flood" in Figure 21.4 (page 170). Clearly, there is some overlap with the anomaly. It is appropriate, therefore, to seek a causal relationship between the anomaly and the Flood.

[14]Roger Y. Anderson, J. Platt Bradbury, Walter E. Dean and Minze Stuiver, "Chronology of Elk Lake sediments: Coring, sampling, and time-series construction," *Elk Lake, Minnesota: Evidence for Rapid Climate Change in the North-Central United States*, ed. J. Platt Bradbury and Walter E. Dean (Boulder: The Geological Society of America, Inc., 1993), 41.

A Bit of a Puzzle

Curiously, however, the Flood seems coincident with the transition *out* of the anomalous period. The Flood is naturally expected to *cause* anomalies, not to *cure* them. Furthermore, the transition out of this anomalous interval seems to mark a transition from wetter to drier conditions, based on the nature of the individual layers, and this is also in the opposite direction of what is expected from the Flood.

The immediate impression is that the anomaly at Elk Lake has nothing to do with the Flood and that the Flood is simply not present at Elk Lake.

But this immediate impression is altogether wrong. A totally different conclusion presents itself when closer attention is given to the anomalous 600 layers themselves.

Solving the Puzzle

I pointed out above that the data from the Elk Lake cores during this 600-layer interval do not seem to paint a coherent picture. I asked how it would be possible for a single lake to go on showing characteristics of three very different stages all at the same time for 600 years.

The answer to this perplexing question is really very simple, if somewhat perplexing itself. It isn't possible. No natural process, not even a Flood of planetary dimensions, can bring about 600 years of the sort of self-contradictory conditions to which the Elk Lake data appear to testify during this anomalous interval.

The best brief summary of the amassed data from Elk Lake that I can give of these 600 layers is that they seem to call for a sudden transition, within a decade or two, from dry wind-swept prairie to moist, still, semi-desert. This moist semi-desert episode apparently persisted with hardly a breeze and never a thunderstorm for 600 years before suddenly reverting back to dry wind-swept prairie.

But this is an impossible picture. Increasing moisture does not convert prairie to semi-desert; it converts prairie to forest. And how can the wind be turned off for 600 years in any region or climate? And how is moisture to be increased while storminess is decreased? This picture just doesn't make sense. I judge it impossible. The created world simply doesn't behave in so irrational and inexplicable a manner.

Now please don't jump to the conclusion that I am saying that something supernatural happened at Elk Lake for 600 years, for I am most certainly not saying any such thing. I am not saying the mix of *data* which is found at Elk Lake during this anomalous interval is inherently impossible. And I certainly don't mean to question the reality of the 600 layers which the researchers identified in the core. Impossibility only enters in when an effort is made to interpret these 600 layers in terms of 600 *years*. They seem to paint an entirely coherent picture when interpreted *as a unit of sediment which was deposited in a single year.*

I suggest that the assumption that the 600 layers which comprise this interval are *annual* is incorrect. I propose that rather than recording 600 years during which 1 to 2 millimeters of sediment were deposited each year, this anomalous section of core should be interpreted as recording a single year during which nearly a meter of layered sediment was deposited on the bottom of Elk Lake.

Corroborating Evidence

The evidence that these 600 layers do not correspond to 600 years is significant. For example, radiocarbon measurements show significantly fewer calendar years spanning this anomalous section of the core than the direct count of layers indicates.

This can be seen as follows, using data and results presented in Appendix C. Lines 9 and 11 of Table C.1 (page 326) show the published radiocarbon ages for annual layers on either side of these 600 layers. For layer 2731 (modern lake stage) the radiocarbon age is 3510±90 years B.P., and for layer 5654 (prairie lake stage) it is 5290±100 years B.P. As explained in the appendix, the old carbon contribution for the prairie lake stage is not known. However, it seems unlikely that it would differ by more than one or two hundred years from the modern lake stage. For the present calculation, it is adequate to approximate the old carbon contribution for the prairie lake stage with the value computed for the modern lake stage from the Appendix C analysis (Equation C.1, page 328) of 605 years. Subtracting an old carbon age of 605 years in both cases and calibrating the residual radiocarbon ages using CALIB rev3.0.3 yields calendar dates of 1092±163 B.C. and 3489±138 B.C. Thus the calendar difference in these two layers computed using these two radiocarbon measurements is (3489±138 - 1092±163 =) 2397±214 years.

The layer number difference is (5654 - 2731 =) 2923, but it is not appropriate to use this difference in the present calculation because, as has already been discussed, 17.6% of annual layers were missed in the layer counting process. Excluding the 600 layers in question from this 17.6% readjustment yields a corrected layer count difference of $(1.176\times(2923-600) + 600 =)$ 3332 layers.

Thus radiocarbon shows (3332 - 2397±214 =) 935±214 (1σ) fewer calendar years spanning this anomalous section of the core than the direct count of layers indicates. Interpreting these 600 layers as the cumulative deposit of a single year brings the layer count into basic harmony with the radiocarbon measurements.

In addition, if these 600 layers were deposited over the course of 600 years, then they would be expected to show a normal pollen accumulation rate, but if they were all laid down in a single year—especially in a single year during which local vegetation was under water for a significant portion of that year—then pollen accumulation rates calculated on the assumption of 600 years might be expected to come out rather low (provided these layers were not all composed of reworked older sediments containing normal pollen concentrations).

In point of fact, "the pollen-accumulation rates are lowest for the period from 5400 to 4800 varve yr."[15] That is, the pollen accumulation rates were lowest in this anomalous 600-layer interval. It appears that pollen accumulation rates which averaged[16] around 20,000 grains/cm^2/yr and fluctuated up to 48,050 grains/cm^2/yr for the prairie period dropped to just 1870 grains/cm^2/yr during this 600 "year" interval.[17]

This observation is difficult to explain if these 600 layers are truly annual. Whitlock *et al.* have suggested that this "may represent a time when slopes were less vegetated than before or after."[18] If there were

[15]Cathy Whitlock, Patrick J. Bartlein, and William A. Watts, "Vegetation history of Elk Lake," *Elk Lake, Minnesota: Evidence for Rapid Climate Change in the North-Central United States*, ed. J. Platt Bradbury and Walter E. Dean (Boulder: The Geological Society of America, Inc., 1993), 259.

[16]J. Platt Bradbury, Walter E. Dean, and Roger Y. Anderson, "Holocene climatic and limnologic history of the north-central United States as recorded in the varved sediments of Elk Lake, Minnesota: A synthesis," *Elk Lake, Minnesota: Evidence for Rapid Climate Change in the North-Central United States*, ed. J. Platt Bradbury and Walter E. Dean (Boulder: The Geological Society of America, Inc., 1993), 312 (graph).

[17]Cathy Whitlock, Patrick J. Bartlein, and William A. Watts, "Vegetation history of Elk Lake," *Elk Lake, Minnesota: Evidence for Rapid Climate Change in the North-Central United States*, ed. J. Platt Bradbury and Walter E. Dean (Boulder: The Geological Society of America, Inc., 1993), 256.

[18]Cathy Whitlock, Patrick J. Bartlein, and William A. Watts, "Vegetation history

fewer plants, there would be less pollen. But why would slopes be less vegetated during this period which is uniformly regarded as more moist than the rest of the prairie period? And if the slopes were less vegetated, and therefore the ground more exposed, how would it be possible to have 600 years without even a single thick layer from any sort of soil erosion during this more moist interval?

Furthermore, if these 600 layers are annual, and represent a change to more moist conditions, some significant change in vegetation, especially toward a higher percentage of trees, would be expected during these 600 years. If, on the other hand, these 600 layers were all laid down in a single year, no change in vegetation would be expected.

The fact that no significant change in the type of vegetation is seen in the pollen record has already been shown above. How is it possible to increase the moisture in a region for 600 years and produce no significant change in its vegetation? Notice that a very significant change in vegetation did accompany the onset of more moist conditions at the transition from prairie to modern lake stages. As Anderson observed above for the prairie to modern transition, "The new expression of moisture and seasonality brought pine, hardwoods, and forest soils to the Elk Lake drainage."[19]

Still further, the transitions from post-glacial to prairie, and prairie to modern lake stages are visibly gradual (Figure 21.4, page 170). If these anomalous 600 layers represent 600 years of climate change, then the transitions into and out of this interval would naturally be expected also to be gradual. But if these layers represent a single year's deposition, gradual transitions at the boundaries are not expected. Figure 21.4 (page 170) clearly shows that the 600-layer interval in question begins and ends abruptly.

What natural climatic factors could account for such abrupt transitions at both ends of a 600-year climate fluctuation? According to Bradbury *et al.*, the "concomitant reduction of clastic indicators sug-

---

of Elk Lake," *Elk Lake, Minnesota: Evidence for Rapid Climate Change in the North-Central United States*, ed. J. Platt Bradbury and Walter E. Dean (Boulder: The Geological Society of America, Inc., 1993), 259.

[19]Roger Y. Anderson, "The varve chronometer in Elk Lake: Record of climatic variability and evidence for solar-geomagnetic-^{14}C-climate connection," *Elk Lake, Minnesota: Evidence for Rapid Climate Change in the North-Central United States*, ed. J. Platt Bradbury and Walter E. Dean (Boulder: The Geological Society of America, Inc., 1993), 46.

gests that climate during this time was unusually calm."[20] There would be no dust blown into the lake if there were no wind, of course. But how is the wind to be shut off over relatively open prairie for 600 years? And why would the wind shut off so suddenly and completely, remain off so long, and resume so suddenly again? Computer climate simulations of the Elk Lake region presented by Bartlein and Whitlock[21] give no hint of such an unusually calm interval.

Bradbury *et al.* go on to suggest "the rapid response is consistent with the triggering of dust suspension at a shear threshold that is determined by soil moisture."[22] According to this explanation, the ground moisture rose sufficiently to inhibit dust production for 600 years. But what actually happens when moisture increases is already known—a gradual transition from prairie to modern lake stages lasting several centuries develops, as is clearly shown by the actual data of Figure 21.4 (page 170), not an abrupt transition at a shear threshold.

The conclusion that these 600 layers do not represent 600 years seems unavoidable.

Conclusion

Interpreting these 600 layers as the stratified deposit of a single year, and correcting the layer count by roughly 17.6% as discussed in Appendix C (page 325), yields the chronology of the Elk Lake core data shown in Figure 21.5.[23] Given this view of these data, it seems clear that Noah's

---

[20] J. Platt Bradbury, Walter E. Dean, and Roger Y. Anderson, "Holocene climatic and limnologic history of the north-central United States as recorded in the varved sediments of Elk Lake, Minnesota: A synthesis," *Elk Lake, Minnesota: Evidence for Rapid Climate Change in the North-Central United States*, ed. J. Platt Bradbury and Walter E. Dean (Boulder: The Geological Society of America, Inc., 1993), 318.

[21] Patrick J. Bartlein and Cathy Whitlock, "Paleoclimatic interpretation of the Elk Lake pollen record," *Elk Lake, Minnesota: Evidence for Rapid Climate Change in the North-Central United States*, ed. J. Platt Bradbury and Walter E. Dean (Boulder: The Geological Society of America, Inc., 1993), 275–293.

[22] J. Platt Bradbury, Walter E. Dean, and Roger Y. Anderson, "Holocene climatic and limnologic history of the north-central United States as recorded in the varved sediments of Elk Lake, Minnesota: A synthesis," *Elk Lake, Minnesota: Evidence for Rapid Climate Change in the North-Central United States*, ed. J. Platt Bradbury and Walter E. Dean (Boulder: The Geological Society of America, Inc., 1993), 319.

[23] The 3σ uncertainty in this chronology is roughly ± 100 years at the time of the Flood. The 17.6% correction, derived from the Modern lake stage, likely overestimates true age by an unknown and increasing amount moving back through the Prairie Period.

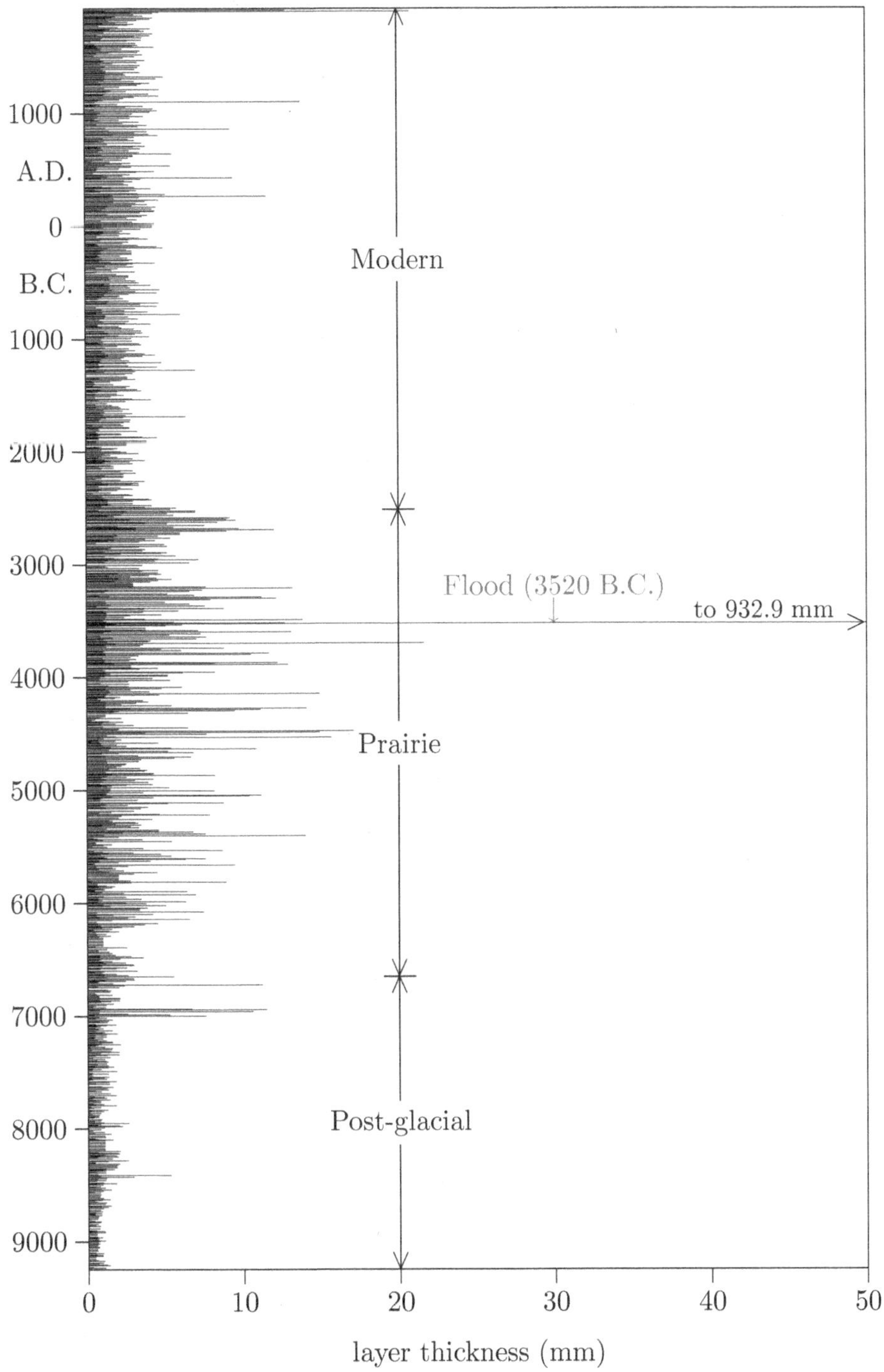

Figure 21.5: Radiocarbon-controlled chronology of the Elk Lake annual sedimentary layer thickness.

Flood was felt at Elk Lake. A 932.9-millimeter-thick annual layer in the midst of 11,000 annual layers just 2 millimeters thick on average is evidence of something remarkably unusual. A thick sedimentary layer at the time of the Flood is the most fundamental expectation of the Flood's action at Elk Lake. When the fact that this thick layer occurs at just the right date for the Flood is added to this expectation, the conclusion that the thick layer was, in fact, produced by Noah's Flood seems unavoidable.

Noah's Flood extended also to Elk Lake, Minnesota.

CHAPTER TWENTY-TWO

CHECK: CYSTS AT ELK LAKE

It is possible to check the conclusion of the previous chapter in a fairly simple way. The check results from the fact that the significant-increase-in-moisture theory and the Flood theory for the origin of the anomalous 600 layers at Elk Lake yield more or less opposite predictions regarding the salinity of Elk Lake during the 600-layer interval. The significant-increase-in-moisture theory leads to an expectation of decreasing lake salinity during the 600-layer interval relative to the previous and subsequent dryer Prairie Period. Contrasting with this, the Flood theory leads to an expectation of increased salinity during the 600-layer interval, because the waters of such a vast flood would necessarily be mainly comprised of ocean water.

Is there a way of determining the relative saltiness of the water of the lake in the past?

Yes. Cysts.

Cysts

A cyst consists of a thin shell made of silica, the major constituent of sand (Figure 22.1)[1]. Cysts are produced by certain types of algae (nonvascular and often single-celled plants). Different species of algae produce cysts which differ from one another in size and appearance. Cysts serve algae

[1]Barbara A. Zeeb and John P. Smol, "Postglacial chrysophycean cyst record from Elk Lake, Minnesota," *Elk Lake, Minnesota: Evidence for Rapid Climate Change in the North-Central United States*, ed. J. Platt Bradbury and Walter E. Dean (Boulder: The Geological Society of America, Inc., 1993), 239–249.

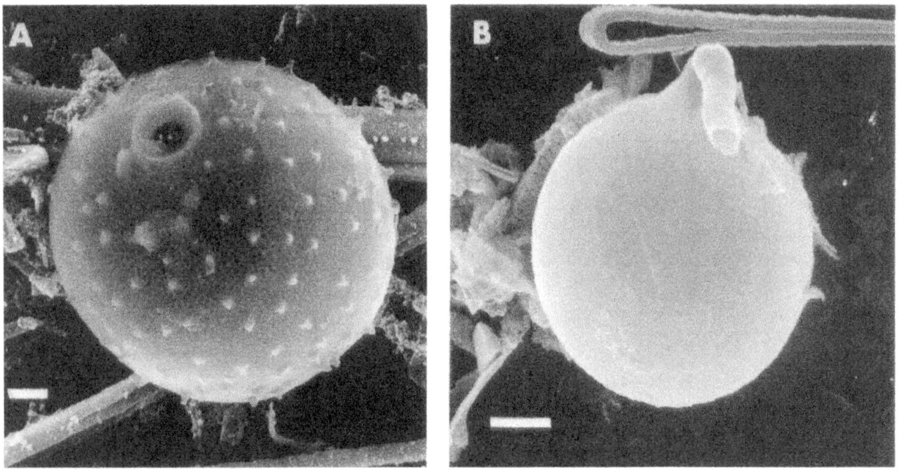

Figure 22.1: Examples of cysts obtained from Elk Lake sediments by Barbara Zeeb and John Smol. (See footnote 1 for the complete reference with additional examples.) Photos were obtained using a scanning electron microscope. The scale bars are 2 microns (79 millionths of an inch).

as a protective "house" in which to live for a period of time. Eventually, the alga cell re-emerges from the cyst through a small opening called a pore.

The importance of cysts to the present study lies in the fact that algae are sensitive to their environment. Species of algae differ in their respective tolerances to various environmental conditions. For example, some types of algae prefer cooler temperatures while other types do better in warmer conditions. This selective sensitivity to environmental conditions presents the possibility of using cysts recovered from Elk Lake's ancient sediments to learn about conditions in the lake in the past.

Salinity

In the present context, the salinity of Elk Lake in the past is of particular interest. Elk Lake is normally a freshwater lake. The Flood would have inundated Elk Lake with ocean water. Since ocean water is salty, it has a higher density than fresh water. Thus it may confidently be predicted that Elk Lake's fresh water was displaced by saline ocean water at the time of the Flood.

It seems likely that Elk Lake would have remained unusually salty for a long time following the Flood. To rid the lake of excess salt requires dilution and flushing with fresh water. But the lake experienced a dry prairie climate for a millennium following the Flood (Figure 21.5, page 183), so the fresh water contribution to the lake back at that time was probably limited. Furthermore, since salt water is denser than fresh water, stratification of the lake following the Flood is probable. Cold, salty water would naturally tend to settle into the deepest portions of the lake. Mixing of these deep saline waters with overlying fresh water would not have been easily accomplished.

In summary, the Flood thesis predicts a sudden influx of salt water to the Elk Lake basin synchronous with the anomalous 600 layers, and persistent elevated saltiness for a significant length of time following the anomalous layers.

A very different prediction follows from the significant-increase-in-moisture theory. Increasing moisture through change in climate would be expected to *decrease* the salinity of Elk Lake. There are only two ways to increase soil moisture (so dust stops blowing into the lake): (1) increase precipitation, or (2) decrease evaporation. In either case, lakes in the region should have become less salty.

Temperature

The past temperature of the lake is also of some interest. The significant-increase-in-moisture theory makes no prediction regarding temperature during the anomalous 600 layers, but the Flood theory predicts that the lake water should have become very cold. This results from the fact that, while the top few meters of the oceans may be quite warm, most of the water of the oceans is deep ocean water, and deep ocean water is only a few degrees above freezing (typically 1–4°C).[2]

Experimental

Zeeb and Smol have painstakingly counted and categorized well over 30,000 cysts from 105 sediment levels at Elk Lake.[3] They have analyzed

[2]George L. Pickard and William J. Emery, *Descriptive Physical Oceanography: an Introduction*, 4th (SI) enlarged edition (New York: Pergamon Press, 1982), 41–42.

[3]Barbara A. Zeeb and John P. Smol, "Postglacial chrysophycean cyst record from Elk Lake, Minnesota," *Elk Lake, Minnesota: Evidence for Rapid Climate Change in the North-Central United States*, ed. J. Platt Bradbury and Walter E. Dean (Boulder:

their numerical data using a statistical method known as "principle components analysis" (PCA). This method separates out independent causes of observed (collinear) variability in a dataset.

Zeeb and Smol's PCA diagram for the interval from 10,400 laminated sediment years ago to 3100 laminated sediment years ago is shown in Figure 22.2. Zeeb and Smol caution that, due to the newness of the whole field of cyst analysis, they cannot be certain which environmental variable is responsible for the variations of their dataset which the PCA analysis has separated out and which they have plotted as "PCA axis 1" and "PCA axis 2." Nonetheless, by comparing their PCA analysis with what is known of the climatic history of Elk Lake, they suggest that PCA axis 1 corresponds to temperature and PCA axis 2 corresponds to salinity, as shown in Figure 22.2.

PCA axis 2 is labeled "wet" and "dry" in conformity with Zeeb and Smol's original figure. This refers to climate around the lake, not conditions in the lake itself. The algae, which live in the lake, are always wet, of course. They are not sensitive to wet and dry climate directly, but rather to salinity of the water in which they live, which increases when the climate is dry and decreases when it is wet. When referring to the lake itself, rather than climate, PCA axis 2 should be labeled "saline" in place of dry, and "fresh" in place of wet. I have added these labels to the right hand axis for ease of reference in the present discussion.

Results

Figure 22.2 starts the lake out (double square in PCA diagram) in a cold, fresh condition 10.4 ky (10,400 years) ago. This is as expected. Cold, fresh water is expected within the newly formed lake following retreat of the glaciers toward the north and east.

This condition persisted, according to the PCA diagram (Figure 22.2), until about 8500 years ago. This was the Post-glacial lake stage. It is set off by the dashed arc at lower right in the PCA diagram.

The lake then warmed, and began to increase in salinity. This is in agreement with conditions expected of the dry Prairie stage, which the climate around the lake had now entered.

The lake then slowly reduced salinity while continuing to warm from about 7400 to 5600 years ago. This is the regime set off by the dashed

The Geological Society of America, Inc., 1993), 239–249.

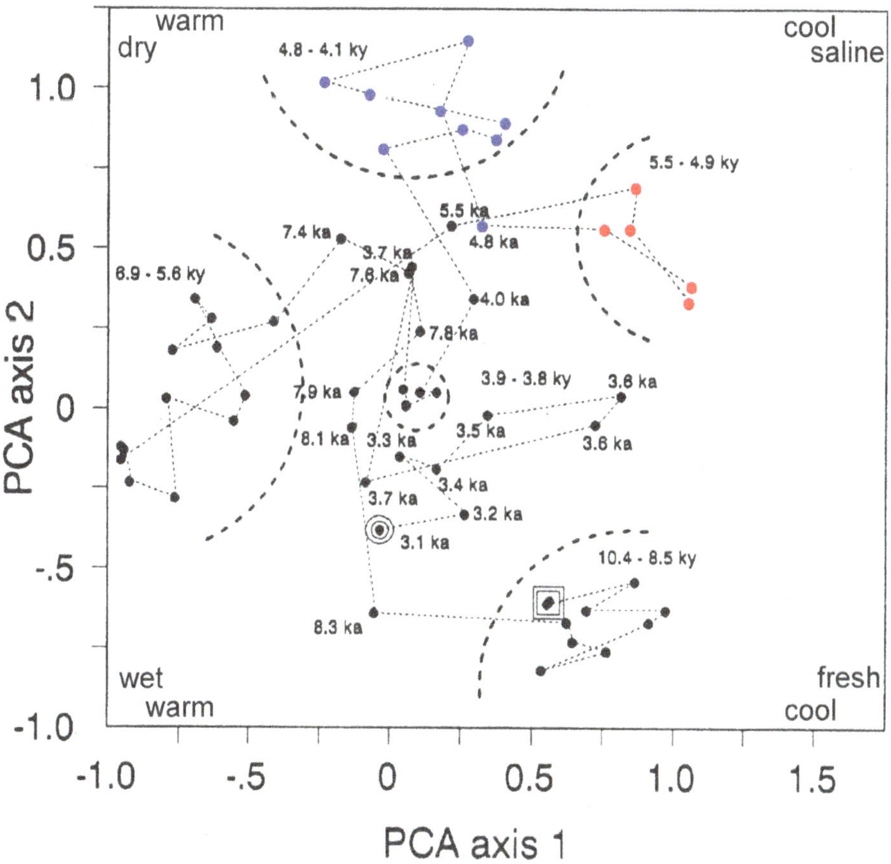

Figure 22.2: PCA diagram for cyst data obtained from Elk Lake sediments. (After Barbara Zeeb and John Smol; see footnote 1 on page 185 for complete reference.) PCA axis 1 explains 40% and PCA axis 2 explains 18% of the total variance. Each plotted point corresponds to a cyst assemblage at a unique depth in Elk Lake's sediments. The development of the lake in terms of temperature and salinity can be traced in time by following the dashed lines from point to point. Time references are given for many points, based on counts of sedimentary layers. The heavy dashed arcs separate out combinations of temperature and salinity which persisted through a significant depth of sediment.

arc at the left of the PCA diagram.

The net trend from the inception of the lake to this point in time was significant warming and moderate increase in salinity.

And then the lake suddenly transitioned out of the warm, moderate salinity regime and into a cold, saline regime, which I have indicated with red dots in the Figure 22.2 diagram. Significantly, the red dots all date to the anomalous 600-layer interval.

Significant-Increase-In-Moisture Theory

The prediction of the significant-increase-in-moisture theory (that the climate was more moist during the anomalous layers interval) fails entirely. The PCA diagram shows an unequivocal transition to drier (more saline) conditions during this anomaly, in exact contradiction to the significant-increase-in-moisture theory.'

Something highly unusual happened at Elk Lake during this anomaly, sending it into an unprecedented temperature – salinity regime. This is most easily seen in Figure 22.3, which includes all of Zeeb and Smol's data from the birth of the lake to relatively modern times. The heavy dashed line in the figure shows that the data points between 5.5 ka and 4.1 ka are in a separate region unto themselves. In Zeeb and Smol's words: "It is interesting that cyst assemblages from 5.5 to 4.1 ky are unique, and have no other analogues during the lake's history."[4] It is clear that the lake experienced something far more unusual than a simple increase in moisture at this time.

Flood Theory

While the PCA results are in blatant contradiction to the significant-increase-in-moisture theory, they strongly corroborate the Flood theory for the origin of the anomalous sediments. The lake not only became suddenly saline, it also became suddenly cold at the time of the anomaly—both of which are necessary if deep ocean water flooded Elk Lake basin, as previously discussed. Following the anomalous sediments interval (i.e., following the Flood), the PCA diagram shows (blue dots) that the lake

[4]Barbara A. Zeeb and John P. Smol, "Postglacial chrysophycean cyst record from Elk Lake, Minnesota," *Elk Lake, Minnesota: Evidence for Rapid Climate Change in the North-Central United States*, ed. J. Platt Bradbury and Walter E. Dean (Boulder: The Geological Society of America, Inc., 1993), 245.

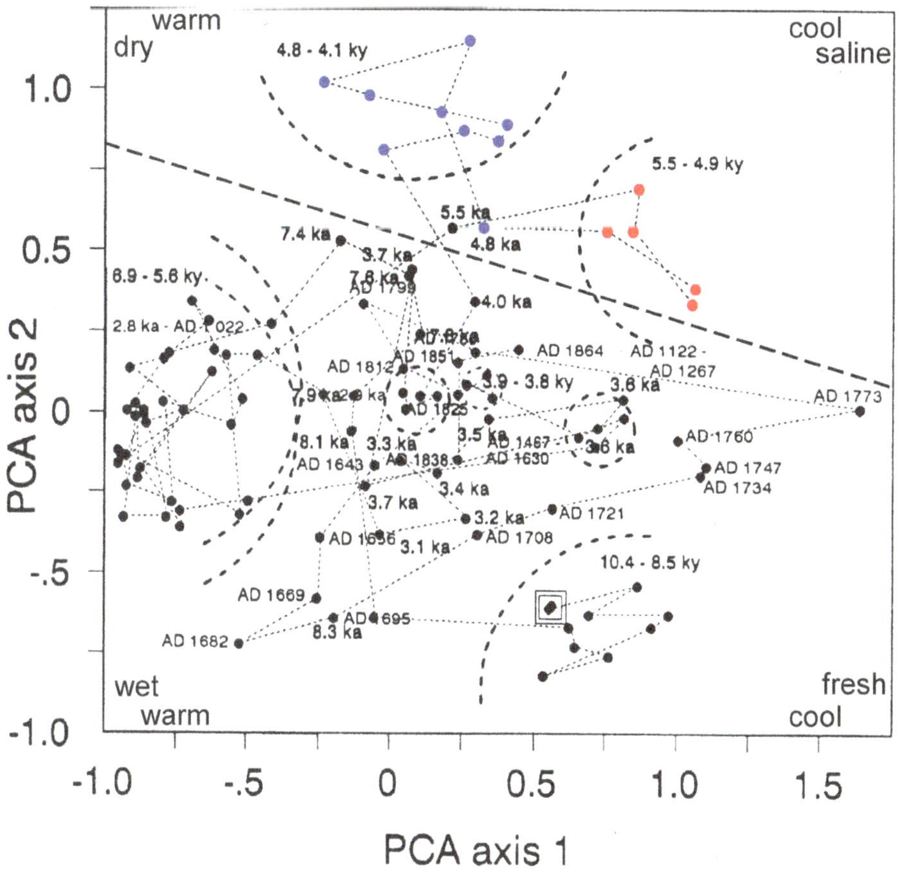

Figure 22.3: Composite PCA diagram for entire cyst dataset obtained from Elk Lake sediments by Barbara Zeeb and John Smol. (See footnote 1 on page 185 for complete reference.)

quickly warmed to normal temperatures again, as expected. And, as already anticipated in the discussion above, high salinity persisted in the lake for an extended time—roughly 700 years.

Conclusion

Zeeb and Smol's cyst data say that Elk Lake became suddenly cold and salty within dating uncertainties of Noah's Flood. Their data contradict the significant-increase-in-moisture theory and corroborate the Flood theory for the origin of the anomalous 600 layers at Elk Lake.

Thus it checks. The claim that Noah's Flood extended to Elk Lake, Minnesota, is corroborated.

Chapter Twenty-Three

Objection: Noah's Flood is Missing at Oyster Pond

Evidences of the Flood have now been demonstrated to be present at widely scattered locations in earth's Northern Hemisphere, including Elk Lake in Minnesota. This seems to imply that Noah's Flood covered at least the Northern Hemisphere. But Neev and Emery object that no evidence of the Flood is present at Oyster Pond in Massachusetts.

Chapter 9 dealt with Neev and Emery's claim that the Dead Sea had not been filled by Noah's Flood. Unfortunately, Neev and Emery did not stop with just the two assertions regarding old shorelines and dilution of the brine at the Dead Sea in their attack on the historical reliability of the biblical Flood narrative in *The Destruction of Sodom, Gomorrah, and Jericho.* They added the assertion:[1]

> Confirmation of the absence of a world-encircling flood during Late Holocene [i.e., anywhere near the time Noah's Flood is usually expected] is provided by studies of sediments in other lakes of the earth—for example, a coastal lake of Late Pleistocene glacial origin in Massachusetts in the United States, where flora, fauna [i.e., plant and animal life], and $\delta^{13}C$ and $\delta^{18}O$ [i.e., carbon and oxygen isotope ratios] show no changes that could have been expected from intrusion of seawater at the biblical date of Noah's Flood (Emery, 1969).

[1]David Neev and K. O. Emery, *The Destruction of Sodom, Gomorrah, and Jericho,* (New York: Oxford University Press, 1995), 120.

It would certainly be curious for the Flood to be present in South Mesopotamia, Palestine, Ireland, the Canadian Arctic, and Minnesota, but missing from Oyster Pond in Massachusetts. (See Figure 23.1; O marks the location of Oyster Pond.) So this objection must now be investigated.

Is the Flood Missing at Oyster Pond?

The "(Emery, 1969)" reference is to a book entitled *A Coastal Pond: Studied by Oceanographic Methods.*[2] It is a slender little book, packed with scientific data of all sorts regarding Oyster Pond in Massachusetts. It contains, for example, graphs and discussion on the rate of escape of methane bubbles from pond sediments. If you find abstract scientific measurements stimulating, then this is the book for you. But if you are inquiring seriously into the historicity of Noah's Flood, I recommend you look elsewhere. This little book contains nothing to allow an intelligent conclusion on that particular question. Worse yet, Oyster Pond hardly recommends itself even for investigation of the question.

Noah's Flood has just been discussed in relation to data obtained from Elk Lake, Minnesota. The thing which made this discussion possible in the case of Elk Lake was that Elk Lake's bottom sediments show clear annual layering. These layers show that the sediments have not been mixed (by fish or clams or worms). As a result, they permit a detailed chronology of the sediments to be worked out. This guarantees that whatever Noah's Flood may have done to the sediments at Elk Lake has been preserved, rather than mixed together with the sediments from before or after the Flood. Furthermore, it permits the portion of the sediment column corresponding to the Flood to be located, because the sediments themselves provide a reasonable chronology, and the date of Noah's Flood is known (i.e., 3520±21 B.C.).

This nearly essential property—annually layered sediments—does not exist at Oyster Pond.

I would not go so far as to say that Neev and Emery's assertion for Oyster Pond is intrinsically impossible, but I do want to point out that substantiation of their assertion demands that somebody have invested a

[2]K. O. Emery, *A Coastal Pond: Studied by Oceanographic Methods*, (New York: Elsevier, 1969).

Figure 23.1: View of Earth from above the North Pole, showing the location of Oyster Pond relative to other locations, previously discussed, in which evidence of the Flood has been found. (M = Mesopotamia, P = Palestine, I = Ireland, D = Devon Island, E = Ellesmere Island, L = Elk Lake, O = Oyster Pond; figure adapted from *Reader's Digest Great World Atlas*, 1963.)

prodigious amount of time and money in the project. Obviously, the most basic requirement, if it is to be shown that Noah's Flood is not represented at Oyster Pond, is an adequate chronology of the pond sediments to be able to tell in which portion of the thirteen meters of sediment at the bottom of the pond evidence of the Flood should be expected. In the absence of annual layering, establishing such a chronology is a major undertaking. I would wish any scientists about to embark on such a project: abundant funding, good luck, and long life—they will likely need all three. Unfortunately, there is no indication Neev or Emery or anyone else has done anything approaching the necessary work to meet even this most basic requirement.

What, *quantitatively*, would Noah's Flood have done to the "flora, fauna, and $\delta^{13}C$ and $\delta^{18}O$" records at Oyster Pond? Neev and Emery provide no answer. But if Neev and Emery have made no quantitative estimate of the probable effects of Noah's Flood on these parameters at Oyster Pond, how can they conclude those effects are absent there? And where are their measurements of flora or fauna or $\delta^{13}C$ or $\delta^{18}O$ versus time at Oyster Pond? There is a low resolution pollen count versus depth of sediment graph, with two uncalibrated radiocarbon dates providing the only chronological indicators in a span of more than ten thousand years in "(Emery, 1969)"[3], but this is woefully inadequate to the present purpose. Beyond this, I have been unable to find anything even remotely to the point in "(Emery, 1969)."

Conclusion

Neev and Emery's assertion that Oyster Pond confirms "the absence of a world-encircling flood" seems, at the very least, premature. They have written a conclusion to an experiment which they have not yet performed. Oyster Pond may have the potential of making some contribution to our knowledge of Noah's Flood, but there seems to be no way to avoid the fact that a whole lot of measurements must be made which have not yet been made before it can possibly make that contribution.

---

[3]K. O. Emery, *A Coastal Pond: Studied by Oceanographic Methods*, (New York: Elsevier, 1969), 25 (Figure 15).

CHAPTER TWENTY-FOUR

A HEMISPHERICAL FLOOD

It has now been found that the Flood extended from southern Iraq to northern Minnesota (Figure 21.2, page 165). Thus the presence of the Flood in the Northern Hemisphere is strongly attested. The obvious next question is, "Did the Flood extend also into the Southern Hemisphere?"

The answer to this question appears to be a pretty solid no. Evidence of action of Flood waters anywhere in the Southern Hemisphere is uniformly missing. This includes, most conspicuously, the vast Antarctic ice sheet. Numerous ice cores have been taken from this ice sheet during the past several decades. None shows any evidence of either topside or bottom-side meltback by the Flood that I have been able to find.

It appears that the waters of the Flood were active in the Northern Hemisphere only; they did not reach into the Southern Hemisphere.

Whence the Water?

This immediately answers several difficult questions which have frequently been asked of the Genesis narrative of the Flood. To begin with, and perhaps most obviously, it answers where the water of the Flood came from and where it went to when the Flood was over. If the Flood occurred in the Northern Hemisphere only, then it is possible for the source, and the eventual sink, of the Flood water to have been the oceans of the Southern Hemisphere.

If you look at a map of the earth, or better still, a globe, you will see that the Southern Hemisphere is dominated by ocean, while the Northern Hemisphere has a relatively high proportion of land. Evidently, quite a

Flood can be gotten up in the Northern Hemisphere if the waters of the southern oceans are relocated to the Northern Hemisphere for a period of time.

Kangaroos and Similar Problems

Another problem which arises within a global Flood context but which does not arise within a northern hemispherical Flood context is how the kangaroos got from Australia to Asia to board the ark and then found their way back home again from the Ararat region after the Flood.

Fossils of kangaroos are evidently found only in Australia. This seems to establish Australia as their only home both before and after the Flood. If it is assumed that the Flood was global and that only those land animals which were aboard the ark with Noah survived the Flood, then kangaroos must have been with Noah aboard the ark. But this raises a difficult problem. How did kangaroos cross the ocean twice between Australia and Asia at the time of the Flood when they appear never to have succeeded in crossing it even once ever before or since?

Actually, the kangaroos are only a small example of a large class of problems of this type arising from the field of zoogeography. This class of problems is not limited to Australia, but it is best illustrated by that continent's odd and oddly assorted animals. Whitcomb and Morris, in their book *The Genesis Flood*, advocating a global cataclysmic Flood, explain:[1]

> The marsupials of Australia consist of very distinct types which find their parallels among the placental animals. For example, there are marsupial *moles*, marsupial *anteaters*, marsupial *mice*, marsupial squirrels (flying phalangers), marsupial *sloths* (koalas), marsupial *gophers* (wombats), marsupial *cats* (dasyures), marsupial *wolves* (thylacines), marsupial *monkeys*, marsupial *badgers* (Tasmanian devils), strange *lizard-like* marsupials called bandicoots, and the *rabbit-like* kangaroos and wallabies. In addition, Australia boasts the only monotremes (egg-laying mammals) in the world: the duck-billed platypus and the spiny anteater.

[1]John C. Whitcomb, Jr. and Henry M. Morris, *The Genesis Flood* (Philadelphia: The Presbyterian and Reformed Publishing Company, 1961), 81.

> On the assumption that the animals of the present world trace their ancestry back to those within the ark, how can we explain the facts that these marsupials and monotremes are found nowhere in the world except in Australia and that the placentals never succeeded in reaching that sub-continent?

Whitcomb and Morris spend several pages discussing this problem, and then conclude:[2]

> The more we study the fascinating story of animal distribution around the earth, the more convinced we have become that this vast river of variegated life forms, moving ever outward from the Asiatic mainland, across the continents and seas, has not been a chance and haphazard phenomenon. Instead, we see the hand of God guiding and directing these creatures...

This, of course, is an admission by Whitcomb and Morris that they have been unable to find any adequate answer to this problem for their global cataclysmic Flood model within the sphere of natural phenomena. And indeed, the problem of the geographical distribution of the marsupial mammals appears solvable only by an appeal to (extra-biblical) miracle within a global Flood context.

To help drive the nature of the problem home, consider the koalas. These Australian marsupials are slow-moving, defenseless, and almost completely arboreal.[3] They feed exclusively on the leaves and buds of the eucalyptus tree. On the assumption of a global Flood, how did the koalas make the journey after the Flood from the ark in Turkey to Australia? Since eucalyptus trees are not native to Asia, what did the koalas eat on this long journey? How did they protect themselves from the carnivorous placentals of Asia? How did they cross the ocean between Asia and Australia? And how did they manage such a trip while placental squirrels and monkeys, for example, which also live in trees and do not suffer the severe limitations of speed, defense, and diet of the koalas, were entirely unable to reach that subcontinent after the Flood?

---

[2]John C. Whitcomb, Jr. and Henry M. Morris, *The Genesis Flood* (Philadelphia: The Presbyterian and Reformed Publishing Company, 1961), 86.

[3]James H. McGregor, "koala," *The Encyclopedia Americana*, vol. 16 (Chicago: Rand McNally, 1962) 502.

Australia is just one region of the globe posing difficulties for a global Flood model. Central and South America constitute another such region. This region holds a monopoly on the native world population of sloths, for example. Two distinct groups of sloths are found there—three-toed and two-toed—with species and subspecies. But no sloths of any sort are found anywhere else in the world. How did *all* the sloths wind up in Central and South America? And how did these creatures, which "are strictly arboreal and do not live outside forest areas" survive the long journey from the Ararat region to South America in a world which had, according to Whitcomb and Morris, just been denuded of all its forests by a global cataclysmic Flood? Note that these sloths "move at an average speed of 14 feet a minute [about one mile in 6.3 hours] on the ground" and that they "sleep 18 out of 24 hours."[4] It takes a long time to travel from the Ararat region to South America (a journey of some 10,000 miles by way of Bering Strait) at a speed of one mile a day!

In the hemispherical Flood model, these difficulties do not arise. According to the hemispherical Flood model, flooding was largely confined to the Northern Hemisphere of the globe. Australia was not flooded, and neither was South America. Their unique faunas had no need to seek shelter in the ark, and no need to make the long journey home from Ararat. They were preserved in their native habitats.

Conclusion

The Flood was restricted to just the Northern Hemisphere of the earth. It appears to have been caused by water from the southern oceans flowing up into the Northern Hemisphere.

How such a thing is possible must, of course, be answered. The solution to this problem will be undertaken, quantitatively, in the next part of this book. Before going there, however, it is necessary to pause and deal with another objection. The objection this time is that Genesis teaches that the Flood was global in extent.

[4]Harold E. Anthony, "sloth," *The Encyclopedia Americana*, vol. 25 (Chicago: Rand McNally, 1962) 100.

Chapter Twenty-Five

Objection: Genesis Teaches That the Flood Was Global

Does Genesis teach that the Flood was global?

I have taken some pains, earlier in this book, to show that the Genesis account of the Flood should be taken at face value. It gives every impression of being an eyewitness account of the phenomena it describes. From the perspective of an objective scientist, the Genesis account of the Flood appears to be, in effect, field notes of Noah's experience and observations of a most extraordinary, singular event. From this perspective, to disregard or explain away these first-hand observations because of their seeming strangeness is pure folly. Our task is not to dismiss or tone down these observations to make them fit more comfortably with our prejudices and limited experience. Our task is to bring our theories about the past into line with these eyewitness observations from Genesis.

Having thus argued, it would clearly be hypocritical were I to advance the idea of a hemispherical Flood should Genesis teach that the Flood was global.

But Genesis does not teach that the Flood was global. Proper scholarship applied to the Genesis narrative of the Flood allows two views in regard to the extent of the Flood. One view is that the Flood was global. The second view is that the Flood was of monstrous proportions, but not necessarily global. Neither of these two views—these two interpretations—can be proven right on the basis of the biblical text alone. To choose between them requires application of extra-biblical data. And when extra-biblical data are applied to them, it is found that the second interpretation is correct, and the first (global) interpretation is incorrect.

The Case for a Global Flood Interpretation

The case for a global Flood interpretation of Genesis has been made by Carol Johnson as clearly, succinctly, and forcefully as I have seen anywhere.[1]

> Yes, it could be from Noah's perspective that all the high mountains were covered, when there were others beyond his vision that were not, but there are other statements in the narrative that were made by God himself that indicate that the Flood was global, since these statements are all inclusive!
>
> To say that the Flood covered only the Northern Hemisphere means that almost the entire continent of South America, half of Africa, and all of Australia and Antarctica would not have been covered with water. How then could God declare, *"And behold I, even I, do bring a flood of water upon the earth, to destroy all flesh, wherein is the breath of life, from under heaven; and everything that is in the earth shall die."* [Genesis 6:17]? I would presume that God meant what He said when He said "all flesh" and "everything" and that this would include the flora and fauna of Africa, South America, and Australia—including kangaroos! This would have to be true according to the text unless there were no animals or people in those continents at the time. ...
>
> As if the statement in Genesis 6 were not enough, God says in Genesis 7:4 "... and every living substance that I have made will I destroy from the face of the earth." Clearly stating the results of the Flood in Genesis 7:21–22, He says, "And all flesh died that moved upon the earth, both of fowl, and of cattle, and of beast, and of every creeping thing that creepeth upon the earth, and every man: All in whose nostrils was the breath of life, of all that was in the dry land, died."

---

[1] Carol Johnson, "Reader's Write," *The Biblical Chronologist* 4.1 (January/February 1998): 8–9. www.BiblicalChronologist.org.

> God states in Genesis 7:23 that "... every living substance was destroyed which was upon the face of the ground ... and only Noah remained alive, and they that were with him in the ark."
>
> Those statements seem to plainly say that every animal was killed by the Flood. That would have to include Africa, South America, Australia, and Antarctica.

The Case for a Non-Global Flood Interpretation

The case for a global Flood interpretation, just presented, rests upon interpretation of the "all"s and "every"s of Genesis 6 and 7 as meaning "all inclusive" so that none whatsoever is left out. This is a possible interpretation, but it is not the only possible interpretation. This is most easily seen by comparison with other Bible passages in which similar expressions are used.

The classic example here is Genesis 41:56–57, in reference to Joseph's famine.

> When the famine was spread over all the face of the earth, then Joseph opened all the storehouses, and sold to the Egyptians; and the famine was severe in the land of Egypt. And the people of all the earth came to Egypt to buy grain from Joseph, because the famine was severe in all the earth.

This passage repeatedly says, "all the earth." Taken literally, it seems to be an assertion that Joseph's famine was a global phenomenon. That means it would have extended to Australia, the Americas, and into the Arctic.

The passage clearly and explicitly says, "And the people of all the earth came to Egypt to buy grain from Joseph." Did Australian aborigines, American Indians, and Arctic Eskimos go to Egypt to purchase grain from Joseph during the seven years of famine?

Here again, the narrative gives every indication of being historical and should, therefore, be regarded as primary observational data from which to understand the phenomenon of the famine which it describes. So I am predisposed to accept that Eskimos paddled their kayaks across the Atlantic ocean, the Mediterranean sea, and up the Nile river if that is what Genesis here says. But I know of no Bible scholar who would suggest that that is what Genesis intends by these "all"s.

The hermeneutical lesson which these verses illustrate is that it is possible for language to be used in Genesis which appears to modern English-speaking people to mean "all inclusive," but which must have been understood by the original audience to mean only "extraordinary largeness."

Genesis 41:56–57 opens the door to the possibility that language used in Genesis 6 and 7, which appears global (or "all inclusive") to us today, may, in fact, be referring to something which, while of stupendous proportions (or, of "extraordinary largeness"), was not global.

Choosing the Right Interpretation

There are thus seen to be two possible interpretations of the Genesis Flood narrative in regard to the geographical extent of the Flood. One view demands a global Flood. The other view allows that possibility but does not demand it.

To the best of my knowledge, there is no rational way to choose confidently between these two interpretations based on the biblical text alone. There are hints that the global Flood interpretation may be incorrect within the text of the Bible itself. For example, Genesis 6:4 records the existence of a genetic line, distinct from Noah, before the Flood, called the "Nephilim," and Numbers 13:33 records the continued existence of this genetic line following the Flood. But both references to the Nephilim are very brief and can be regarded only as a hint, not a proof.

Whitcomb and Morris have attempted to choose between these two interpretations, in favor of a global Flood, in their book, *The Genesis Flood*, but I think their effort must be judged a failure.

They advance three points in this regard. The first is that "most universal terms [in the Bible] are to be interpreted literally."[2] I concede this point, but I fail to see how it helps choose confidently between the two possibilities in the specific case in question. Is this specific case one of the majority, or is it one of the minority?

---

[2]John C. Whitcomb, Jr. and Henry M. Morris, *The Genesis Flood* (Philadelphia: The Presbyterian and Reformed Publishing Company, 1961), 56.

Their second point is that "the context determines the meaning."[3] They clarify what they mean by this in this specific case by quoting M. M. Kalisch and italicizing for emphasis "*the universality does not lie in the words merely, but in the tenor of the whole narrative.*"[4]

The statement that "context determines meaning" is a sound hermeneutical principle, but it cannot be applied in the sense of Kalisch's quote in this instance. To do so is to commit the logical error called begging the question.

The whole question under consideration is "Does Genesis teach that the Flood was global?" The "yes" side has advanced the argument that the "all"s and "every"s of Genesis 6 and 7 show that the Bible does teach that the Flood was global. The "no" side has advanced a counter-example; they have shown a case in Genesis where a lot of "all"s similar to those in Genesis 6 and 7 don't equate to global. Whitcomb and Morris reply for the "yes" side. They say that the counter-example does not apply to the "all"s in Genesis 6 and 7 because context determines meaning, and the whole context of Genesis 6 and 7 is global, showing that the "all"s of Genesis 6 and 7 are meant to be understood globally. But that is begging the question. To say that the context (i.e., "the tenor of the whole narrative") of Genesis 6 and 7 is global is just a way of asserting that Genesis teaches the Flood was global. But this cannot be asserted (as a premise) since it is, in fact, the question which is to be resolved. Obviously, the "no" side does not grant the premise that the whole context of Genesis 6 and 7 is global. That is what the "yes" side must *prove* if they wish to convince the "no" side.

Whitcomb and Morris state their third point as follows [original emphasis throughout]:[5]

> But our third and most impelling reason for interpreting the universal terms of Genesis 6–9 literally is that *the physical phenomena* described in those chapters would be quite inconceivable if the Flood had been confined to one section of the earth. While it would be entirely possible for a seven-

[3]John C. Whitcomb, Jr. and Henry M. Morris, *The Genesis Flood* (Philadelphia: The Presbyterian and Reformed Publishing Company, 1961), 56.

[4]John C. Whitcomb, Jr. and Henry M. Morris, *The Genesis Flood* (Philadelphia: The Presbyterian and Reformed Publishing Company, 1961), 57.

[5]John C. Whitcomb, Jr. and Henry M. Morris, *The Genesis Flood* (Philadelphia: The Presbyterian and Reformed Publishing Company, 1961), 60.

> year famine to have gripped the Near East without at the same time affecting Australia and America (cf. Gen. 41:57), it would *not* have been possible for water to cover even *one* high mountain in the Near East without inundating Australia and America too!

Notice, first of all, that this is an appeal to extra-biblical data. The text of the Bible does not teach about the alleged behavior of water under the influence of gravity to which this argument appeals. We learn about how water behaves under the influence of gravity from our own real life experience, and we learn it more thoroughly and accurately through the systematic study of real-life experience we call science. So Whitcomb and Morris, in common with everybody else, have been unable to settle this interpretive question on the basis of the text of the Bible alone. Clearly, the idea of a global Flood is not something which Genesis, in and of itself, teaches.

I will show, in the next part of this book, that Whitcomb and Morris's grasp of the science which they have advanced to settle this interpretive question is faulty. According to the laws of physics (in particular, the law of gravity), it *is* in fact possible for water to cover mountains in the Near East without inundating Australia. Thus their third point also falls to the ground.

Conclusion

There appears to be no rational way to choose confidently between the global ("all inclusive") and non-global ("extraordinary largeness") interpretations of Genesis 6 and 7 based on the biblical text alone. Both interpretations are possible if the question is confined to biblical material only.

But only one interpretation can be true. The Flood was either global in extent, or it was not. One of these two interpretations must be false. As long as biblical data are all that are available on this interpretive question, it is best to allow both possibilities to stand, so as not arbitrarily to reject the truth and embrace error.

But, fortunately, the question is no longer restricted to biblical data alone. The date of the Flood is now clear, and the true nature of this historical event has already been significantly clarified through the application of previously gathered scientific data within their proper chronological framework. There is, admittedly, still much to be learned in this process. Nonetheless, the scientific data which have been considered to the present time appear already to be conclusive against the global Flood interpretation of Genesis 6 and 7. It appears to be an interpretive error to insist that Genesis teaches that the Flood was global.

Chapter Twenty-Six

Conclusion to Part 3

Given the modern biblical chronology date of Noah's Flood—3520±21 B.C.—it becomes reasonable to look for evidences of the Flood in various geophysical reservoirs. The prime requirements for such a quest are that the geophysical reservoir under consideration have a reliable secular chronology associated with it, and that this chronology reach back prior to the Flood.

Evidences of the Flood are easily found in the Northern Hemisphere using either data from archaeology or data from suitable geophysical reservoirs. In contrast, evidences of the Flood appear to be lacking for the Southern Hemisphere.

Thus the Flood appears to have been hemispherical in extent. It appears to have covered most or all of the Northern Hemisphere of the earth, but not to have extended much into the Southern Hemisphere.

This conception of the Flood immediately answers the question of where the water for the Flood came from, and it avoids such problems as how Australia's kangaroos, located in Australia both before and after the Flood, survived the Flood.

Remaining to be explained is the mechanism by which the water from the southern oceans was displaced into the Northern Hemisphere and held there for the better part of a year. This topic comprises the next part of this book.

PART 4

THE MECHANISM OF NOAH'S FLOOD

Chapter Twenty-Seven

How to Cause a Hemispherical Flood

Why would the oceans of the Southern Hemisphere have left their basins to join with the oceans of the Northern Hemisphere? What impelled these waters to behave this way? How could the Northern Hemisphere have been kept flooded to great depth for the better part of a year, while the ocean basins of the Southern Hemisphere were dry? Can any explanation of such a thing be found?

In fact there does appear to be a way in which such a thing can be brought about. The explanation is somewhat staggering, though it conforms fully to the known laws of science. To understand it, a brief excursion into the fields of astronomy, physics, and geophysics is necessary.

Tidal Phenomena

The oceans are kept in their place today by the force of gravity. Since the earth is (approximately) a sphere, the force of gravity points radially in toward the center of the earth everywhere. As a result, the surface of the oceans is not flat, but rather is curved into the shape of a sphere also.

The surface of the oceans is routinely disturbed from its spherical shape by the tides. Tides result from the gravitational attraction of the moon and the sun. Tides demonstrate that the surface of the oceans does not need to conform to a perfect sphere. Tides are, in fact, bulges of large geographical extent on the surface of the oceans. In point of fact, the water of the oceans simply obeys the law of gravity. The shape of the surface of the oceans is determined by the relative strengths of the gravitational pull upon the water of whatever massive bodies exist nearby.

The dominant gravitational force on the oceans today is exerted by the earth itself because it is so massive and so close to (just beneath) the oceans. Hence the surface of the oceans is roughly a sphere. But the moon is pretty massive and not too far away, so it exerts an appreciable gravitational pull on the oceans too. And the sun, while very much farther away than the moon, is also very much more massive than the moon, so its gravitational pull on the water is also felt, distorting the surface of the oceans from that of a sphere

While tides are very helpful in illustrating that the shape of the surface of the oceans is determined by gravity and does not need to conform to a sphere, tidal phenomena cannot solve the present problem. The difficulty is that tides always, necessarily produce two bulges at the same time on opposite sides of the earth. (Any elementary discussion of the cause of tides will explain why this is the case.) For the present problem, a phenomenon is needed which produces a single, very large bulge on one side of the earth only.

Remote Gravitational Attractor

But suppose the moon or a comet or some other astronomical object were to pass quite close to the earth. Could the gravitational attraction of such an object create the sort of bulge which is required?

No, a close encounter with some other astronomical body will not do the job either. The difficulty is that both the solid earth and the fluid oceans would experience exactly the same acceleration due to the passing object. The oceans and the solid earth would move together in perfect concert toward the astronomical object. There would be no one-sided bulge.

It may thus be concluded that the Flood was not due to gravitational attraction, tidal or otherwise, between the oceans and any imagined close encounter with some astronomical body. Evidently, the cause of the unusual distribution of the water of the oceans at the time of the Flood must be sought within the earth itself.

The Geoid

This takes the discussion out of the realm of astronomy and into the realm of geophysics. And what is being discussed, in the jargon of the geophysicist, is the shape of the *geoid.*

The geoid is a very simple thing. It is simply the actual shape of the surface of the oceans due to the gravitational attraction of the mass of the earth and its rotation alone—tidal effects are ignored. Geophysicists extend the oceans beyond their shorelines by constructing imaginary canals all through the continents. In this way the geoid can be referenced at every point on the surface of the earth, not just over the oceans.

The foregoing discussion has treated the geoid as if it were a perfect sphere. In actuality, it is not a perfect sphere. It is, first of all, a somewhat flattened sphere, with the radius at the equator greater than that at the poles. This results from the rotation of the earth and the fact that the solid earth is itself flattened in this way. The bulk shape of the earth and the geoid is therefore a spheroid rather than a sphere. This flattening is not very interesting in the present context, and I will ignore it in what follows.

Of much greater interest is the fact that the geoid does not conform to a perfect spheroid either. There are local hills and valleys on the geoid. What this means is that the ocean surface is not all at the same distance from the center of the earth. The ocean surface has hills and valleys like the land, though not nearly as pronounced as mountains and valleys on land. For example, there is a valley in the surface of the ocean of very broad geographical extent which reaches a depth of 100 meters (330 feet) below the spheroid surface just south of India.[1]

Geoid Warps

These hills and valleys—or "warps," as the geophysicists call them—are caused by the fact that the distribution of mass within the earth is not uniform. Some regions are made of denser rock than others. Since the force of gravity is proportional to mass, and the mass of rock per unit volume varies from place to place on the surface of the earth, the force of gravity varies slightly from place to place as well. This produces geoid

[1]George D. Garland, *Introduction to Geophysics*, (Toronto: W. B. Saunders Company, 1979), 164.

warps.

You might suppose that you could slide down the slopes of a geoid warp in a boat, but in fact you can't. If gravity could pull you down the slope of one of these hills or valleys in the surface of the oceans, it could also pull the water down the slope, and in that case the hill or valley would soon disappear. The hills and valleys exist because that is the shape the slightly uneven gravity over the surface of the earth, due to the uneven distribution of mass within the earth, has pulled the water into. Gravity is, in fact, everywhere perpendicular to the actual surface of the oceans (i.e., the force of gravity is everywhere perpendicular to the geoid). There is no component of the force of gravity parallel to the surface to pull you down the slope. And if you were on such a slope in a boat, you would not feel that you were on a slope at all. Because gravity would be perpendicular to the slope, you would feel that the slope was level. This is so because vertical and horizontal are always relative to gravity. As the direction of gravity changes, so do horizontal and vertical. That is why people who live a quarter of the way around the earth from you do not feel like they are walking around sideways.

This is a little foreign to everyday experience, and, therefore, a little difficult to grasp at first. But hills and valleys in the surface of the oceans—geoid warps—really do exist. They are not an imaginary thing of science fiction. They exist because water must obey the law of gravity.

It is important to grasp the reality and basic physical behavior of geoid warps before proceeding because it is about to be shown that a large wooden boat appears to have floated for about half a year on a very large geoid warp, and I don't want you to suppose that it would have slid down the slope of that geoid warp and crashed into the dry ocean bottom in the Southern Hemisphere.

How to Cause a Large Enough Geoid Warp

Now the problem of how a geoid warp large enough to explain the Flood could possibly be produced must be tackled.

I have studied this problem pretty thoroughly, and I can find only one possible solution within the realm of natural phenomena. It requires (unthinkable though it may seem) that the inner core of the earth be displaced.

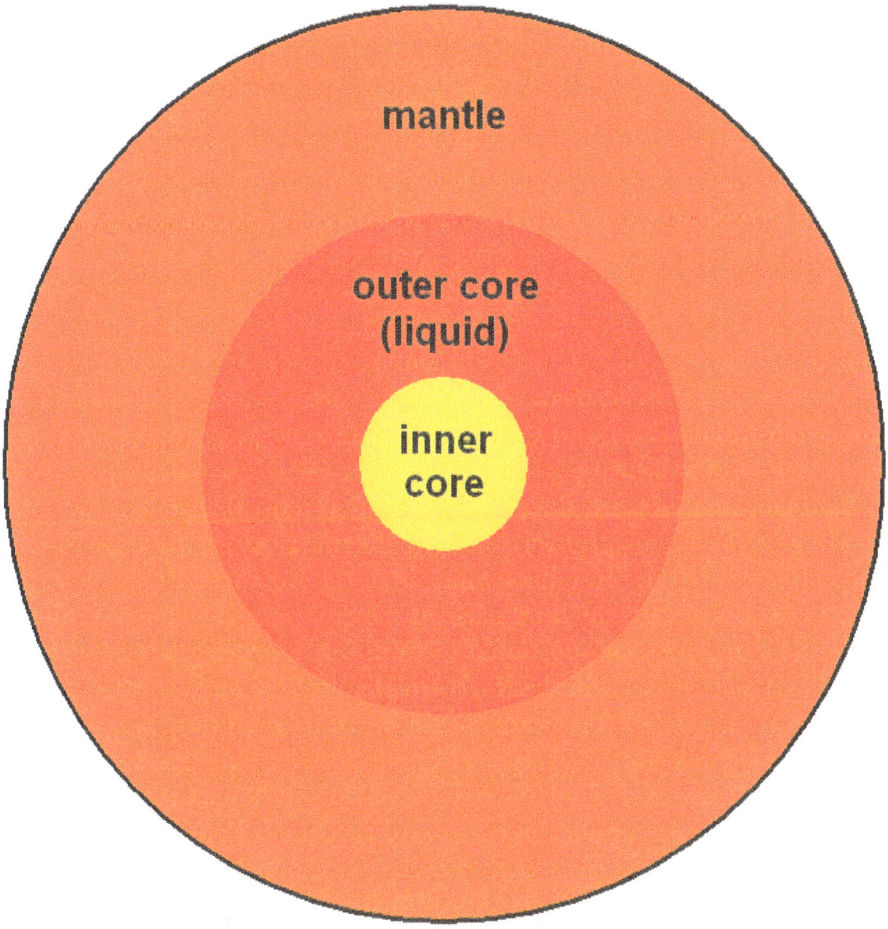

Figure 27.1: Scale cross section of the earth.

Earth's Insides

The earth is made up of layers, somewhat like an egg (Figure 27.1). The part we are most familiar with is the crust (black in the figure). It is like the shell of the egg. It is quite thin compared to the rest of the earth, and it is solid.

Beneath the crust is the mantle. It is also solid. Think of it like the white of a hardboiled egg.

Beneath the mantle is the core. It is like the yolk of the egg. The outer core is liquid, like the yolk of an uncooked egg, and the inner core is solid, like the yolk of a hardboiled egg.

Inner Core Displacement

How the inner core is maintained in its accustomed central position is actually a more complex matter than at first appears. This will become increasingly clear in subsequent chapters. For now, simply notice that the inner core is not rigidly coupled to the mantle in any way. It is surrounded by outer core fluid and is thus free, in principle, to be displaced from its usual central position. The force required to move it is not small, but let us ignore the whole matter of how a displacement of the inner core might be brought about for the time being and concentrate only on what will happen at the surface of the earth when the inner core is displaced.

Said simply, displacement of the inner core toward the North Pole will produce a positive geoid warp (i.e., a hill in the ocean surface) at the North Pole and a negative geoid warp (i.e., a valley) at the South Pole.

Said more elaborately, since the inner core is more dense than the surrounding liquid of the outer core, moving the inner core toward the North Pole will increase the total mass which is in the Northern Hemisphere of the earth and decrease the total mass in the Southern Hemisphere. Since the force of gravity is proportional to total mass, gravity at the surface of the earth will increase in the Northern Hemisphere and decrease in the Southern Hemisphere. The water of the oceans (and everything else on earth) will feel a slight but very real gravitational attraction toward the North Pole. The water, which is free to move, will flow northward.

If the inner core were displaced toward the North Pole, it would seem to an observer on the surface of the earth that the direction of horizontal had changed. Surfaces which were previously level would now seem to be tilting downhill toward the North Pole, even though they had not actually moved at all—the local gravitational field is all that would have changed. Water which had been standing on horizontal surfaces before the inner core had been displaced would flow "down" those surfaces toward the North Pole after the displacement. If you were standing on a beach which ran east and west during such a displacement you would see the water of the ocean begin to flow "up" the beach toward the North Pole. More accurately, you would feel that the ocean had tipped up and that the beach had tipped down so that the water of the ocean was being poured out across the beach.

Noah described it this way:[2] "...all the fountains of the great deep [i.e., the ocean] burst open..."

---

[2]Genesis 7:11; NASB 1975.

CHAPTER TWENTY-EIGHT

CHECK: DEPTH OF NOAH'S FLOOD

The hemispherical Flood model requires the existence of some mechanism by which the southern oceans can be made to flow up into the Northern Hemisphere. I have proposed that displacement of the inner core of the earth from its normal central position into the Northern Hemisphere of the core is the needed mechanism. This mechanism can be checked quantitatively as follows.

First, assume that the inner core rose directly beneath the North Pole. This should be a reasonable first approximation considering the ease with which evidences of the Flood have been able to be identified in the northern latitudes compared with the apparent absence of such evidences in the southern latitudes. Second, use the observational data supplied in Genesis to estimate the depth of the Flood at the point of observation. Then, third, use the well-known equation for the gravitational potential to calculate how deep the water would be at various points on the globe due to displacement of the inner core, and see whether water deep enough to satisfy the biblical narrative can be achieved in this way.

The Biblical Data

The single most major piece of quantitative data regarding the depth obtained by the Flood is found in Genesis 8:4. There it is learned that "the ark rested upon the mountains of Ararat." This observation is made in a context of the waters receding, after having covered "all the high mountains everywhere under the heavens" (Genesis 7:19). This says immediately and unambiguously that the water of the Flood attained a depth sufficient to cover mountains in the Ararat region.

The Mountain

Can the mountain the ark landed on be specified, and thereby pinpoint the location on the globe where this observation was made? Unfortunately, it cannot.

I will return to this question in a short while, showing that the Genesis narrative of the Flood does provide a number of clues which may be used to make a reasonable guess at the actual mountain the ark landed on. But for now, so as not to multiply uncertainties and thereby weaken this first test of the model, I will follow a worst case analysis rather than a most probable analysis.

Tradition

Tradition says the ark landed on Mount Ararat, the tallest peak in modern Turkey (Figure 28.1). But tradition has not proved to be an always reliable guide to ancient biblical sites. And numerous expeditions to Mount Ararat to search for remains of the ark have so far failed to come up with anything convincing—despite some claims to the contrary.[1] For the present purpose, something more substantial than tradition is needed to go on.

Bible

It is commonly believed that the Bible says the ark landed on Mount Ararat. In point of fact, however, it does not do so.

Genesis 8:4 records only that "the ark rested upon the mountains of Ararat." Bible scholars observe that the use of the plural—mountains—precludes the identification of any particular mountain by the Bible writer.[2] Rather, they suggest, "the mountains of Ararat" should be understood as referring to a mountainous area in a region known as Ararat.

[1]For a refutation of one such claim see: R. E. Taylor and Rainer Berger, "The Date of 'Noah's Ark'," *Antiquity* LIV (1980): 34–36.

[2]For example: W. W. Gasque, "Ararat," *The International Standard Bible Encyclopedia*, ed. Geoffrey W. Bromiley (Grand Rapids: Eerdmans, 1979), 233.

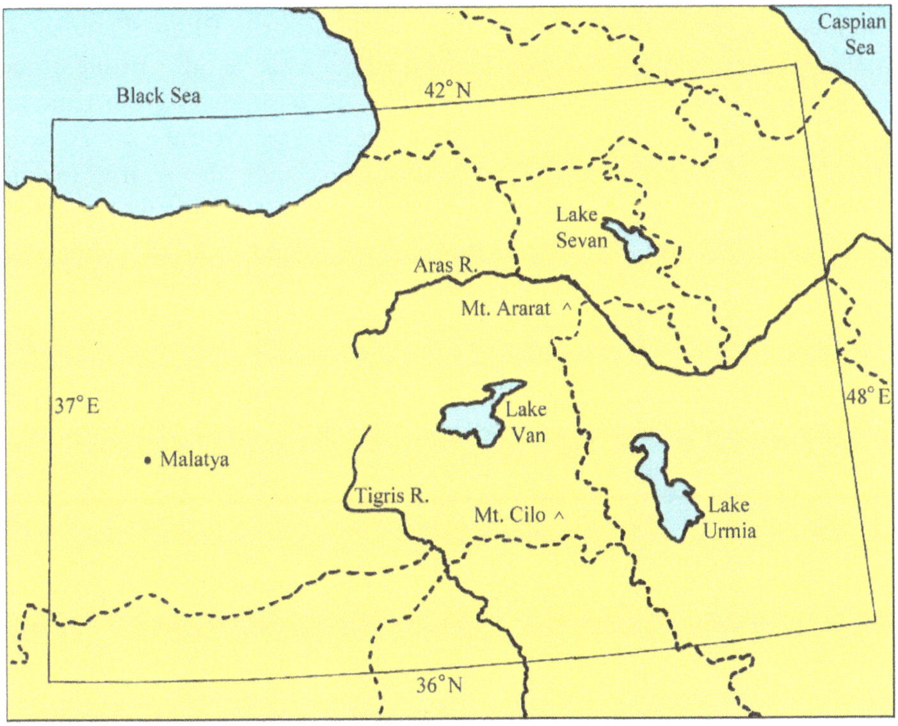

Figure 28.1: Map of Ararat region and its surroundings. Dashed lines follow borders of modern countries.

The Location of the Ararat Region

Where is this region called Ararat to be found on the globe? If the Bible does not specify the mountain the ark landed on, might it nonetheless specify the general area in which the ark came to rest?

"Ararat" is mentioned three times in the Old Testament in addition to Genesis 8:4. Unfortunately, these references are not very helpful in and of themselves when it comes to the problem of locating this region. The first two are parallel accounts in 2 Kings 19:37 and Isaiah 37:38 which mention "the land of Ararat" but do not say where that land is. The third (Jeremiah 51:27) is a prophecy against Babylon which mentions the "kingdoms of Ararat, Minni and Ashkenaz." Here again no specific geographical details are provided.

Apparently, no direct indication of the location of the Ararat region is to be found in the Bible. But these Bible references do help clarify the meaning of the word "Ararat" as it is used in the Bible. They show that "Ararat" is indeed a reference to a region (a "land") rather than a specific mountain. They also show that, at least during the time of Jeremiah (approximately 600 B.C.), a kingdom of Ararat was known. And Genesis 8:4 adds to this the expectation that the Ararat region contained a mountainous area.

Genesis 11:2 provides some potentially important indirect biblical evidence regarding the location of Ararat. It says that the descendants of Noah came into the land of Shinar (which is generally understood to be the plain between the Tigris and Euphrates rivers in Mesopotamia) "as they journeyed east." This would place the region of Ararat to the west of Shinar. Unfortunately, this geographical hint is complicated by the Septuagint rendering of this verse which reads "as they moved from the east." This would put Ararat in the opposite direction, to the east of Shinar.

It seems necessary to conclude that the location of the Ararat region cannot be ascertained with any certainty from the Bible alone.

Biblical Archaeology

The most direct link to the location of the Ararat region seems to come about through a combination of biblical and archaeological data. Specifically, an independent kingdom known to the Assyrians as Urartu was familiar to the Assyrians as a neighbor and enemy from about 850 B.C. Urartu is known from archaeological and historical data to have been located in the vicinity of Lake Van in modern Turkey (Figure 28.1, page 221).

The kingdom of Urartu seems to be the only possible candidate for the kingdom of Ararat which Jeremiah mentions. In addition to this archaeological lead it should be noted that the geographical region associated with Urartu encompasses the traditional Mount Ararat. This says that Christians, for many centuries at least, have, for whatever reasons, held this to be the Ararat region. Furthermore, this region seems to satisfy the biblical evidence the best. It contains many mountains. It is also the region from which the Tigris river begins, making it easy to see how Noah's descendants could have come upon the land of Shinar by following the southeastward course of this river.

Scholarly Consensus

By taking these biblical and archaeological facts into consideration, scholars seem to have come more or less unanimously to the conclusion that the Ararat region corresponds roughly to eastern Turkey today:

> A country in the region of Lake Van in Armenia, where today the borders of Russia, Iran, and Turkey converge.[3]

> A country in [the historical] Armenia, its center being Lake Van.[4]

> The name Ararat, as it appears in the Bible, is the Hebrew equivalent of Urardhu, or Urartu, the Assyro-Babylonian name of a kingdom that flourished between the Aras and the Upper Tigris rivers from the 9th to the 7th century BC.[5]

> Spreading northwestward [from the region of Lake Van] into Transcaucasia, Urartu extended its settlements to the region of Mount Ararat, the valley of Araxes (modern Aras), and Lake Sevan, and westward, for a time, as far as Melitene (modern Malatya).[6]

The general region indicated by these scholars is shown in the map of Figure 28.1 (page 221). The area bounded by longitudes 37°E on the west and 48°E on the east, and latitudes 36°N on the south and 42°N on the north includes Lake Van, Malatya, Lake Sevan, and all of the other landmarks mentioned in the preceding quotes. It seems certain that Noah's ark came to rest somewhere within this bounded area.

---

[3]W. W. Gasque, "Ararat," *The International Standard Bible Encyclopedia*, ed. Geoffrey W. Bromiley (Grand Rapids: Eerdmans, 1979), 233.

[4]H. A. Hoffner, Jr., "Ararat," *The Zondervan Pictorial Encyclopedia of the Bible*, vol. 1, ed. Merrill C. Tenney (Grand Rapids: Zondervan, 1976), 255.

[5]"Ararat, Mount," *The New Encyclopaedia Britannica*, vol. 1 (Chicago: Encyclopaedia Britannica, 1997), 518.

[6]Georges Dumezil, "Armenia," *The Encyclopedia Americana* (Chicago: Rand McNally, 1962), 266.

The Depth

The tallest mountain in this region is Mount Ararat itself. Its elevation is 16,900 feet (5,150 meters) above mean sea level today. Assuming this mountain was within visible range of the ark says immediately, at least in first approximation, that the water of the Flood exceeded this depth.

I say "in first approximation" because it cannot simply be assumed that this was the height of the mountain at the time of the Flood. A mountain which erodes as little as one inch per year will have lost over 450 feet in 5500 years, the time which has elapsed since the Flood. Additional uncertainties on this order arise from questions of the possible subsidence or elevation of the Ararat plateau since the Flood. But these considerations seem unlikely to change this depth estimate by more than about 500 feet (150 meters) in the worst case. So let us add this extra to the present height of the mountain, to yield an estimated Flood depth of 5300 meters in the Ararat region.

Note in passing that this sets the surface of the Flood waters in the Ararat region 5.3 kilometers (3.3 miles) above the surface of the oceans today. Since the average depth of the oceans at present is 3.8 kilometers, this estimate says that the Flood depth in the Ararat region significantly exceeded an average ocean depth above mean sea level.

The Gravitational Potential Calculation

It is possible to calculate fairly accurately how deep the water would be around the globe for various displacements of the inner core using the equation for the gravitational potential. To do this to very high accuracy requires numerical modeling methods which take into account the actual sea floor and continental topography of earth. But such high accuracy is not needed for the present purpose. I have carried out the needed calculations to sufficient accuracy with the aid of a fairly simple computer program (Appendix D).

Figure 28.2 shows the results of my calculation. In this figure I have plotted the depth of water versus degrees of latitude from the North Pole. Thus 0 degrees corresponds to the North Pole, ±90 corresponds to the equator, and ±180 corresponds to the South Pole.

The depth of the water in this figure is relative to the normal ocean surface today. Notice that the vertical scale is greatly exaggerated rela-

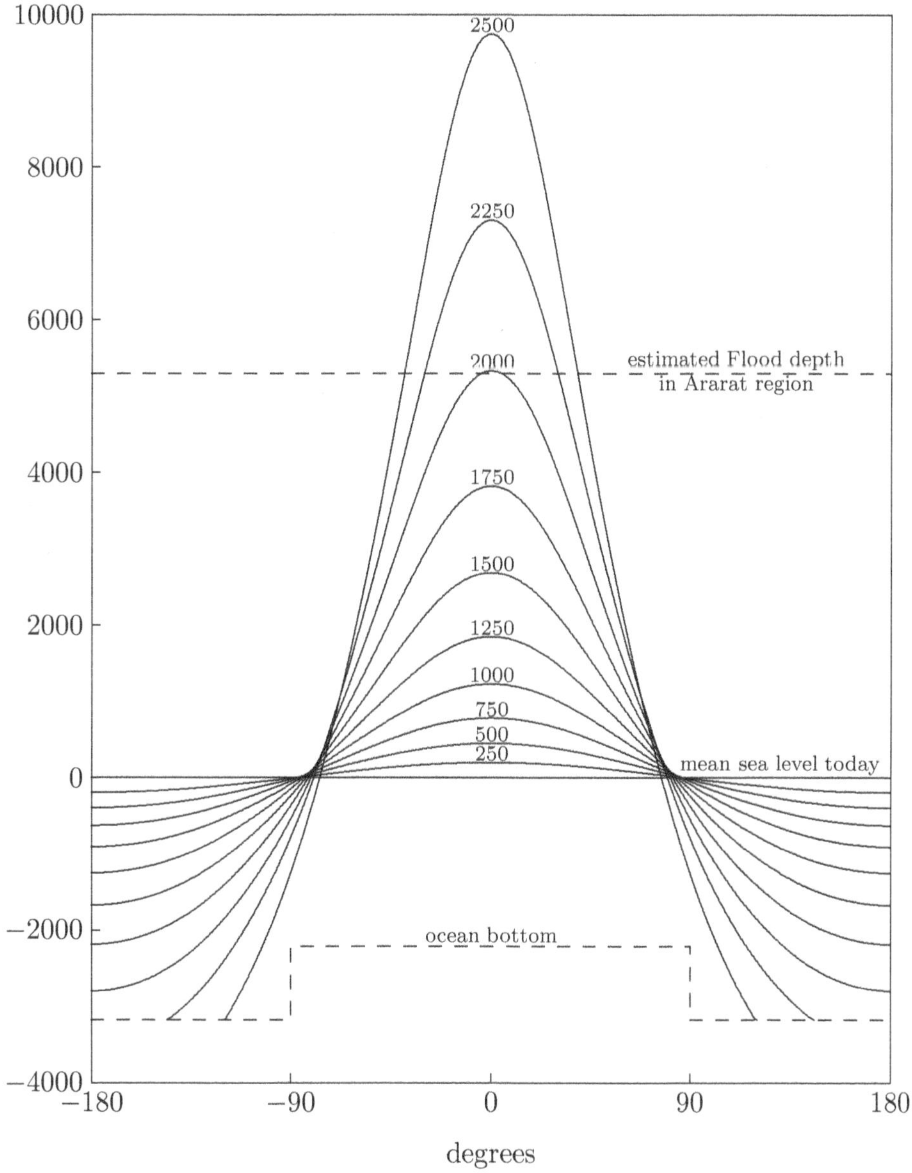

Figure 28.2: Depth (in meters above mean sea level) of the water of the Flood around the globe for various displacements of the inner core. Each depth curve is labeled with the corresponding inner core displacement in units of kilometers.

tive to the horizontal scale. (The circumference of the earth—from -180 to +180 degrees—is about 40 million meters.) This gives the geoid warps shown artificially steep apparent slopes. In actual fact, the steepest slope would have been only about 0.06 degrees, corresponding to an increase in depth of the Flood ocean of about 1 meter per kilometer moving along its surface toward the North Pole.

The dashed line at the bottom of the figure shows the ocean bottom. I have raised the ocean bottom in the Northern Hemisphere relative to the Southern Hemisphere to compensate for the excess continental volume in the Northern Hemisphere relative to the Southern Hemisphere.

The different depth curves shown result from displacements of the inner core from the center of the earth toward the North Pole. They are in steps of 250 kilometers displacement.

The largest free displacement of the inner core shown is 2250 kilometers. At 2258 kilometers (according to the Preliminary Reference Earth Model) the solid inner core contacts the solid mantle—the inner core has hit the ceiling.[7] Notice that a portion of the southern ocean is completely dry at this displacement.

This does not seem to be the largest possible displacement, however. It seems probable that the inner core would be able to embed itself to some extent in the underside of the mantle for a sufficiently large velocity of the inner core relative to the mantle. Thus a displacement to 2500 kilometers is also shown. It results in the geometry shown in the scale diagram of Figure 28.3.

Result

The estimated depth of the Flood in the Ararat region which has been calculated above from the biblical Flood narrative is shown in Figure 28.2 (page 225) by the horizontal dashed line at 5300 meters. It is immediately obvious that a Flood of sufficient depth to satisfy the biblical narrative can be obtained in this way.

[7] Frank D. Stacey, *Physics of the Earth* (Australia: Brookfield Press, 1992), 454–455.

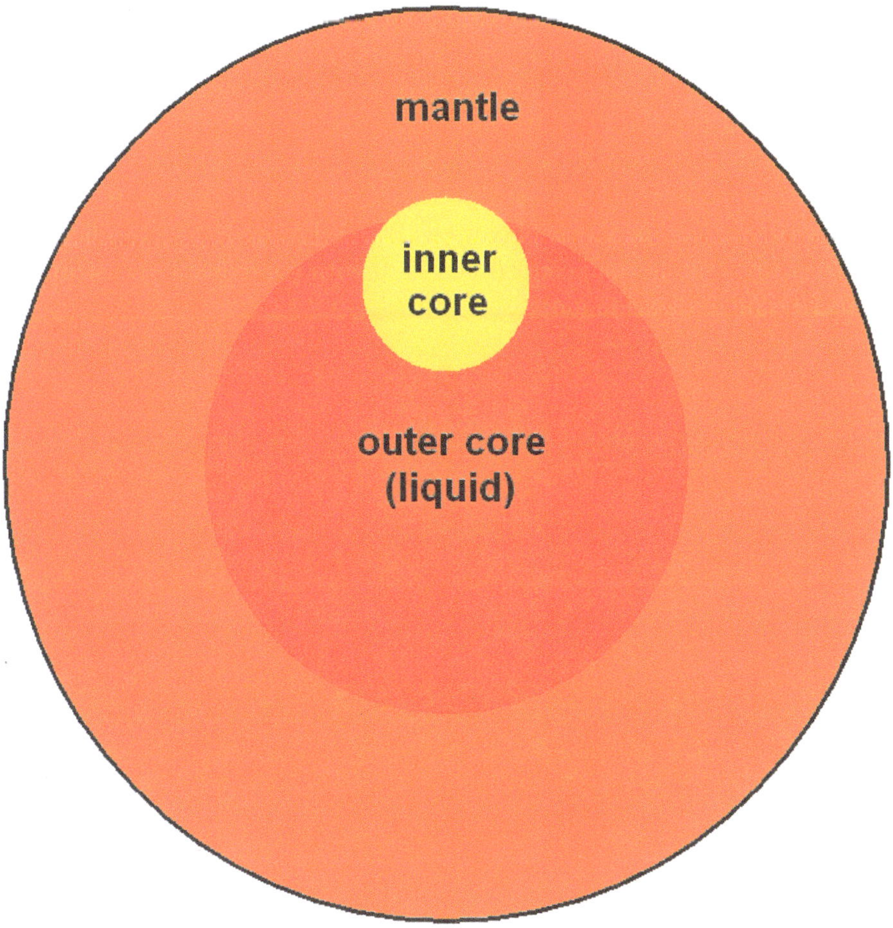

Figure 28.3: Scale cross section of the earth with inner core displaced from center by 2500 kilometers.

Conclusion

Thus it checks. The problem of how a large enough geoid warp could be produced is indeed solved by displacement of the inner core.

Since this is the *only* mechanism which I have been able to find which appears adequate to the task of explaining Noah's observational data, it appears to me that, no matter how incredible it may seem, the inner core of our planet was, in fact, significantly displaced in a northerly direction at the time of the Flood, 5500 years ago, and that this displacement is responsible for the vast flooding observed by Noah. Said another way, Noah's observations of the Flood, recorded in Genesis, appear to inform us that the inner core of our planet can be, and indeed was, significantly displaced from its normal central position, with consequent catastrophic impact upon civilization globally 3520 B.C.

Chapter Twenty-Nine

Check: Behavior of the Atmosphere

It is possible to check the inner-core-displacement mechanism for the Flood in another way. This check is quite simple.

So far, discussion has centered around the water of the Flood. The atmosphere is also of considerable interest.

Noah reports two important atmospheric phenomena in connection with the Flood. The first is that it rained day and night for the first 40 days of the Flood.[1] The second is that the waning of the Flood was accompanied by a dry (rainless) wind.[2] Are these observations congruent with the idea that the Flood was caused by displacement of earth's inner core?

Indeed they are.

The Rise and Fall of the Atmosphere

The atmosphere is made up of air. Air is a gas. Gases and liquids are fluids. They are free to flow in response to gravity.

Because the water of the oceans was not rigidly coupled to the surface of the earth, it was free to flow north in response to the geoid warp caused by the inner core's displacement from center. The air would have behaved in the same way. The air would have piled up deeply in the Northern

[1]Genesis 7:4; Genesis 7:12.

[2]Genesis 8:1–3.

Hemisphere, just as the water did. As a result, the atmosphere in the Northern Hemisphere would have become thicker, while in the Southern Hemisphere it would have become thinner.

To make a thicker atmosphere requires that some air be lifted to higher altitudes than normal. Air cools when it is lifted to higher altitudes. As a result, water condenses out and eventually falls as snow or rain.

This phenomenon is familiar today. In Oregon, for example, the Cascade Mountains form a high barrier to prevailing winds from the Pacific Ocean to the west. Moist air has to rise to higher altitudes to flow over the mountains. In doing so, the air cools and releases its moisture as rain. This gives rise to a temperate rain forest along the coast.

If the inner core moved to the north at the start of the Flood, then its gravitational attraction would have piled up air increasingly in the Northern Hemisphere all the while that it was moving. This would naturally have given rise to rain as the piling air increased in altitude. Thus rain is to be expected in the Northern Hemisphere all the while the inner core was moving northward at the start of the Flood.

The opposite effect would have happened during the waning of the Flood. As the inner core began to return to its normal central position, the geoid warp would have shrunk in size day by day, and the mountain of air which had been piled up for months in the north would have begun to flow back to the south. Since the flow of air in the atmosphere is simply what we call a wind, the waning of the Flood is expected to have been accompanied by a wind in the Northern Hemisphere.

This wind is expected to have been warm and dry. It would have been like air which has crossed the Cascade Mountains today. Coming down the far side of the mountains, it naturally warms as it reduces altitude. Having already lost its moisture when ascending the west side of the mountain range, the result is a warm dry wind and a so-called "rain shadow" region to the east of the mountain range.

During the Flood, the air had lost its moisture as the atmosphere had thickened in the north. The high altitude air in the thick northern atmosphere is expected to have cooled radiatively and thus to have become relatively dense. As the geoid warp began to shrink during the waning of the Flood, this cold, dense, high altitude air would have begun to drain away under gravity, yielding a warm, dry katabatic wind (rather than a warm, dry rain-shadow wind) at the surface in the Northern Hemisphere.

Conclusion

The initial 40 days and nights of rain which Noah recorded are easily explained by the inner-core-displacement mechanism for the Flood. They would have been due to the lifting of air to high altitudes in the north as the inner core journeyed northward.

Notice briefly an important consequence of this insight. Air can move quickly. According to multiple Internet sources, wind speed in the jet stream can reach 200 miles per hour, and the world record measured ground wind speed is 231 miles per hour. At this speed air can move half way around the globe in just 2.25 days. But Noah observed that the rain lasted 40 days, and he makes no mention of hurricane-force winds. Thus it may be concluded that the inner core did not complete its journey northward in just a few days. Rather, it was progressing northward throughout the first 40 days of the Flood.

This conclusion checks with the Genesis narrative of the Flood in two ways. First, it harmonizes with Noah's silence regarding any wind at the start of the Flood. A travel time of 40 days for the inner core moving northward yields a wind speed of just 13 miles per hour for air to flow half way around the globe. This wind speed corresponds to a gentle to moderate breeze—nothing particularly noteworthy. Second, this travel time means that the water depth in the Northern Hemisphere would have been increasing throughout the first 40 days of the Flood. This harmonizes completely with the Genesis record of the Flood previously discussed in Chapter 16.

As another consequence, notice that this answers the question about the geographical extent of the 40 days and nights of rain. Precipitation (whether as rain or snow) would have been a hemispherical phenomenon—active in the Northern Hemisphere, where the atmosphere was thickening, but not in the Southern Hemisphere, where it was thinning.

And finally, the rainless wind which Noah mentions as the first phenomenon accompanying the receding of the water during the waning of the Flood is also easily understood within the context of this Flood mechanism. It would have been due to dry air from the thickened northern atmosphere descending in altitude as the northern atmosphere returned to normal in consequence of the inner core returning back toward center.

In summary, if the Flood was caused by displacement of the inner core, then the waxing of the Flood should have been accompanied by more or less continuous rain in the north, and the waning of the Flood should have been accompanied by a drying wind in the north. Noah records that both of these things were true.

CHAPTER THIRTY

HOW THE INNER CORE WAS DISPLACED

A basic mechanism for the Flood is now in hand—displacement of earth's inner core. This basic mechanism places the hemispherical Flood model on a quantitative basis and immediately explains a great deal about the Flood.

It is now clear, for example, where the water for the Flood came from, and where it went back to following the Flood. Both the source and the sink of the Flood water are no longer mysteries. We also now understand how such incredibly deep water could have been observed by Noah in the Ararat region. Furthermore, it is now clear how it is physically possible for the Flood to have covered high mountains in the Northern Hemisphere without having had to cover all the high mountains over the entire planet. In addition, the sudden, simultaneous extinction of widespread cultures in the Northern Hemisphere 5500 years ago can be understood. The meltback of the ice caps at Devon Island and Ellesmere Island, and the absence of evidence of similar meltback in the Antarctic regions of the globe can be explained. It is now clear why it is no problem for the Flood that kangaroos and koalas, but no squirrels or monkeys, should be found in Australia, and why it is no problem that sloths should be found in Central and South America, and only in Central and South America. And it is now clear why Noah reported that it rained day and night for weeks at the start of the Flood, and that a dry wind accompanied the waning of the Flood. All of these previous mysteries, and more, are solved given the single, simple postulate of displacement of earth's inner core.

But further mysteries remain to be solved.

Topping the list of mysteries at this point is how earth's inner core could possibly have been displaced to the extent demanded by Noah's observational data. This problem must now begin to be tackled. To solve this mystery, two questions need to be answered:

1. Just how far was the inner core displaced from center?

2. By what means was the inner core moved that distance?

Question 1: How Far Did the Inner Core Go?

Figure 30.1 shows the gravitational potential calculation again. It shows that, were the inner core to rise directly beneath the North Pole, a displacement of 2000 kilometers would be needed to get water deep enough to cover Mount Ararat were Mount Ararat situated at the North Pole. But Mount Ararat is not situated at the North Pole. The Ararat region, in which the ark landed and which includes Mount Ararat, is located roughly 52 degrees from the North Pole. This offset is shown by the vertical dashed line in the figure. It shows that, to get deep enough water to cover Mount Ararat at its actual geographical location, were the inner core to rise beneath the North Pole, a displacement of the inner core much greater than 2000 kilometers would be required.

But a displacement of 2000 kilometers is already 89% of the way to contact between the inner core and the mantle. It thus seems likely that collision between the inner core and the mantle did, in fact, occur at the time of the Flood.

This same conclusion seems to follow logically from Noah's observations of the Flood. Having taken 40 days to reach full depth, the Flood persisted at this depth for the next 110 days (Figure 16.3, page 119). When a rock is thrown vertically upward, gravity pulls it back down immediately upon its reaching maximum height. Why did the Flood persist at full depth for 110 days—why did the inner core not begin an immediate descent back to center after reaching its maximum displacement on Day 40? The most obvious answer to this question is that the inner core

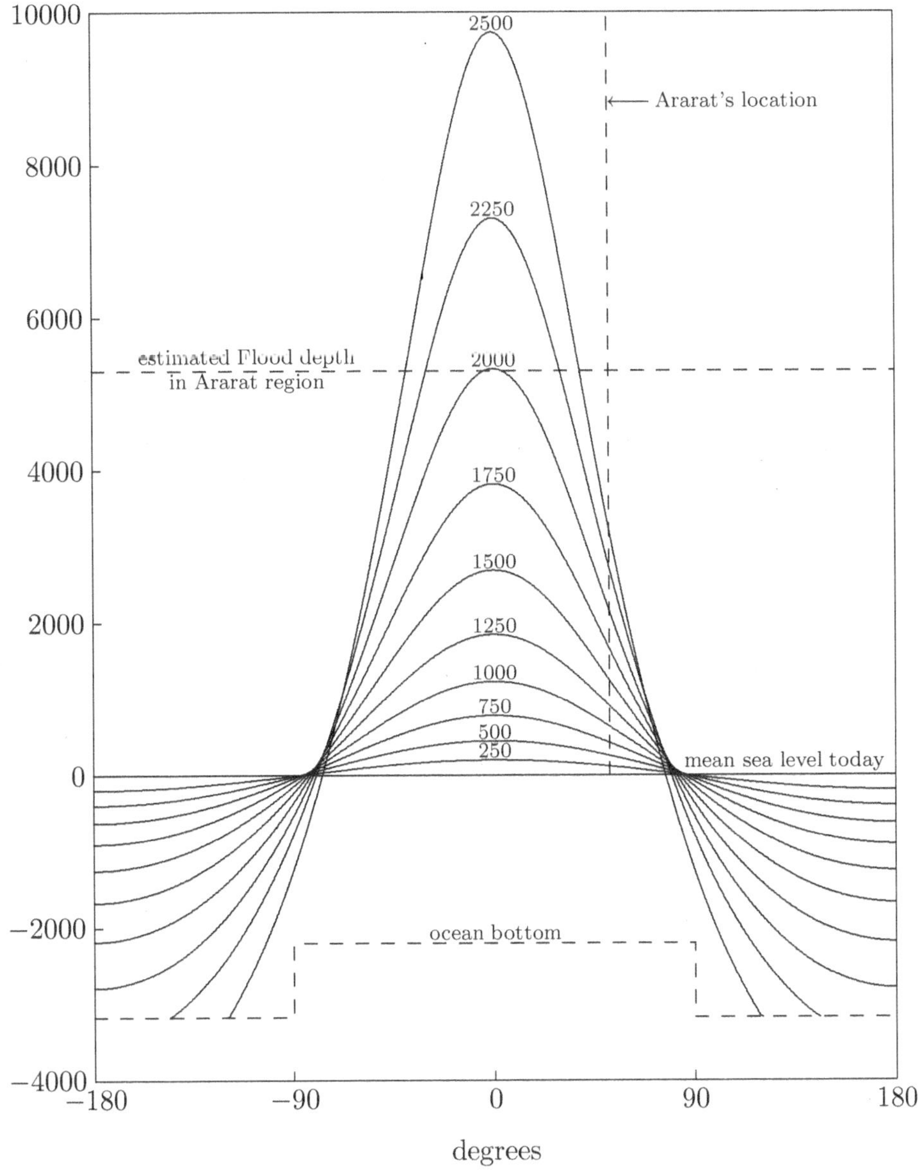

Figure 30.1: Depth of the water of the Flood (in meters above mean sea level) for various displacements of the inner core (in kilometers).

contacted the mantle on Day 40, and that it remained pinned there for the next 110 days.

Thus it appears that the inner core actually collided with the mantle at the time of the Flood. That this is indeed what happened will become increasingly clear in subsequent pages.

Question 2: How Was the Inner Core Displaced?

What force operating on the inner core could have driven it against gravity and through dense, viscous outer core fluid a distance of more than 2250 kilometers from center to collision with the mantle?

I have spent more time and effort trying to answer this question than on anything else associated with the writing of this book. It is not an easy question. After much searching and more than one false start, I suggest the following working hypothesis.

In panoramic overview, the process began with the inner core being nudged off center. (Chapter 32 deals with how this was accomplished.) This initiated a runaway process in the core (discussed below) in which the inner core was expelled from center. The inner core dragged outer core fluid with it as it was driven from center. This contributed to a core-wide toroidal fluid circulation. This circulation assisted in carrying the inner core to the mantle against the increasing gravitational restoring force. (Gravitational acceleration starts out at zero on center and climbs more or less linearly to 7.4 m/s^2 by the time the inner core contacts the mantle.) This circulation held the inner core pinned to the mantle for 110 days (Figure 30.2). Only as it weakened sufficiently did the inner core begin its journey back toward center.

The Runaway Process

Focus now on the runaway process mentioned above by which the inner core was expelled from center.

The outer core normally has no core-wide circulation. The core may be thought of, for the present purpose, as normally motionless, with the inner core motionless on center.

At the time of the Flood, an external force moved the inner core slightly off center (Chapter 32). Imagine the process which took place

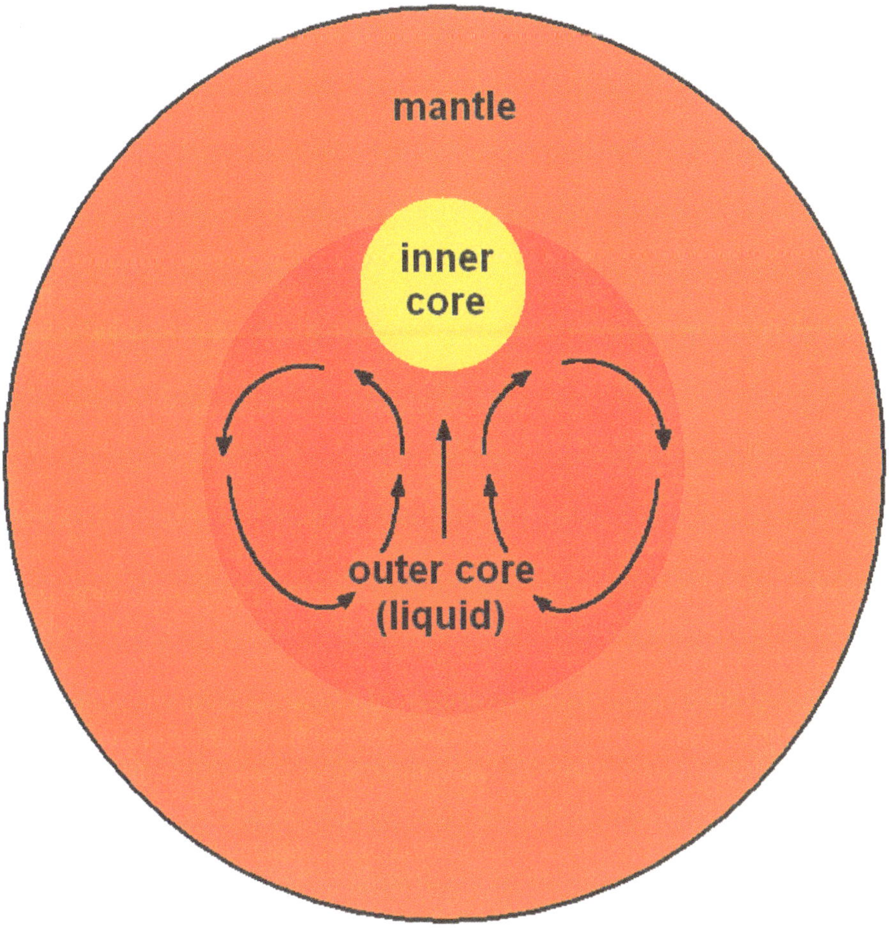

Figure 30.2: Full Flood depth persisted for 110 days, implying that the inner core was held pinned against the mantle by rising outer core fluid currents.

as the inner core began to move off center, rising vertically. Fluid would have flowed in to fill the vacancy which the moving inner core was just beginning to make. Because pressure inside the earth increases radially toward center, this fluid would have been moving from a region of lower pressure into a region of higher pressure. Because outer core fluid is compressible, it would have compressed in this higher-pressure region. The result of this compression would have been shrinkage of the filling fluid's volume. This shrinkage would have made it necessary for additional fluid to flow in to fill up the vacancy. Thus the relatively small initial volume vacated by the inner core would have acted as somewhat of a sink for outer core fluid. Fluid will naturally flow into a sink. But the inner core would have acted as a one-way valve. Fluid would have been able to flow into the sink only from beneath the sink—the solid inner core would have blocked flow from above. Thus the flow into the sink would have been directional.

There appear likely to have been three consequences of the flow into the sink just described, contributing to a runaway ejection of the inner core from center.

First, directional flow into the sink would have promoted a core-wide fluid circulation. Symmetry says that this core-wide circulation would have been predominantly toroidal. Since the inner core would have been at the center of this torus, its displacement from center would have been assisted by this circulation.

Second, fluid falling into the sink would have collided with the inner core, transferring momentum to it and driving it farther from center.

Third, compression of outer core fluid in the central sink region vacated by the inner core would have resulted in a loss of outer core fluid volume. This would have been only partially compensated by decompression of the solid inner core as it moved from higher pressure on center to lower pressure out ultimately in contact with the mantle (Chapter 33). Thus there would have been a net reduction in total core volume. As a result, the whole earth would have shrunk (or contracted, or collapsed) a little. This would have provided the energy (from gravitational potential energy) which drove the runaway process.

As the inner core was moved farther off center, more volume would have been vacated beneath the inner core, further enlarging the sink. More fluid would have flowed in to fill this vacancy, further amplifying the core-wide toroidal circulation and further driving the inner core from center. This runaway process would have continued all the while the region normally occupied by the inner core was filling with outer core fluid.

The overall process may be thought of as a slight (but very real) gravitational collapse of the planet due to the loss from center of the usual inner core "foundation" of the earth.

This collapse would necessarily have been accompanied by rising pressures in the entire core. As matter collapses (or falls) inward, it gains momentum. This momentum would have been halted only by core pressures in excess of those normally present maintaining hydrostatic equilibrium. These enhanced core pressures would have exacerbated compression of outer core fluid, further accelerating the runaway process.

Ultimately, the inner core was halted by collision with the mantle, collapse would subsequently have been halted by elevated core pressures, and a rebound would have begun.

The overall process was ponderous. Noah's observations inform us that the characteristic time scale for collapse and rebound was a year.

Conclusion

The qualitative discussion above constitutes a working hypothesis for how the inner core may have been displaced to the mantle at the time of the Flood. Something nudged the inner core off center (Chapter 32) initiating a runaway ejection of the inner core to the mantle. The loss from center of the solid inner core "foundation" of the earth inevitably brought about a slight, relatively slow (months-long) gravitational collapse of the whole earth. Thus, at the start of the Flood, the earth began a slow contraction inward lasting some months, which was followed by a similarly slow rebound outward.

This working hypothesis is, regrettably, qualitative only. Quantification is needed to correct and refine it as necessary, or to corroborate its basic soundness. Unfortunately, quantification, in this instance, is a major project. The Navier-Stokes equations for a viscous, compressible fluid with moving boundaries must be solved. While this promises to be a rewarding endeavor with potential to teach us much about our planet's core, it is inappropriate to hold up publication of this book to bring about this needed quantification. The soundness of the kernel of the working hypothesis—displacement of the inner core to the mantle and accompanying gravitational collapse of the planet—are strongly corroborated in a number of ways, as subsequent chapters will show. This kernel carries with it the implication that the inner core's normal residence on center is not stable. And this, in turn, implies substantial potential hazard to civilization (discussed in Part VI). There follows an unavoidable ethical mandate to publish as expeditiously as possible at this stage.

CHAPTER THIRTY-ONE

VISCOSITY OF THE OUTER CORE

The working hypothesis for the inner core displacement mechanism of the previous chapter dictates that the outer core have a very high viscosity. This is shown in the present chapter. Very high outer core viscosity may thus be regarded as an integral part of the working hypothesis.

Viscosity

The viscosity of a fluid tells how "thick" or "gooey" it is. Viscosity measures resistance to flow. The lower the viscosity, the higher a fluid's fluidity is. Water has a viscosity of roughly 0.001 Pascal-seconds (Pa-s). Honey has a viscosity over 1000 times greater than water, around 2 Pa-s.

There are some kitchen substances which we might not ordinarily think of as fluids at all because they have relatively high viscosities. Peanut butter is an example. It has a viscosity of roughly 200 Pa-s. Lard is another example. It has a viscosity closer to 2000 Pa-s. These substances will flow if subjected to moderate pressures. We cause peanut butter to flow when we spread it on our bread with a knife.

Tar pitch (or bitumen)—the "glue" holding blacktop road aggregate together—is the most commonly familiar fluid having a really high viscosity. Its viscosity is roughly 2×10^8 Pa-s.

It is possible to find substances having yet higher viscosities. Even the mantle of the earth has a viscosity—roughly 10^{21} Pa-s. This is huge, of course. You cannot spread it with a knife. But at the enormous pressures characteristic of the interior of the earth, mantle rock will slowly deform and flow.

Outer Core Viscosity

The viscosity of outer core fluid may be regarded as currently among the most unknown of scientifically measured quantities.

> Estimates of outer core viscosity span 14 orders of magnitude.[1]

This quote summarizes 39 studies and four different measurement methods. Values range from a minimum of 10^{-3} Pa-s (i.e., similar to water) to a maximum of 8.6×10^{11} Pa-s (i.e., over 1000 times more viscous than tar pitch).

Noah's Observations Pertinent to Outer Core Viscosity

Noah's observations of the Flood enable this unhappy state of affairs to be amended when they are understood in light of the displacement mechanism of the previous chapter. They then strongly corroborate the high end of this measurement range. They do so as follows.

On Day 150, the inner core was about to begin its descent back to center. Thus the force exerted by the outer core fluid currents holding the inner core to the mantle (Figure 30.2, page 237) would have just balanced the force of gravity acting on the inner core. This is the case for which the equation for terminal velocity of a sphere falling in a fluid in a gravitational field applies.

$$v_t = \frac{2R^2 g(\rho_s - \rho)}{9\eta} \tag{31.1}$$

In the present case, R is the radius of the inner core (1.2×10^6 meters) and g is the acceleration due to gravity at 2250 km. The Preliminary Reference Earth Model (PREM)[2] says this is roughly 7 m/s^2. (I am deliberately being approximate with this calculation. I wish to avoid discussion of many fine details which turn out to be inconsequential for the present purpose. Fine details include the fact that, for a displaced inner core, PREM values for g will not be quite right, the fact that the

[1]Secco, R. A. (1995) Viscosity of the Outer Core, in Mineral Physics & Crystallography: A Handbook of Physical Constants (ed T. J. Ahrens), American Geophysical Union, Washington, D. C.. doi: 10.1029/RF002p0218.

[2]Frank D. Stacey, *Physics of the Earth* (Australia: Brookfield Press, 1992), 454–455.

mantle boundary is present disturbing flow around the inner core, the fact that pressures and densities in the core during contraction and subsequent rebound will be higher than PREM values, etc.) Other parameters in the equation are the average density of the inner core, ρ_s, (take this to be 1.3×10^4 kg/m^3 in agreement with PREM), ρ, the average density of the outer core (1.1×10^4 kg/m^3 according to PREM), and η, the viscosity of outer core fluid we wish to solve for.

To solve for η, it is necessary to supply a value for v_t, the terminal velocity. This is the velocity of the fluid currents relative to the inner core. At the start of the Flood, the inner core rose to the mantle (a distance of more than 2250 kilometers) in 40 days. Thus its average speed was roughly 0.65 m/s (1.5 miles per hour). This may be taken as the characteristic velocity in the present case.

Result

Solving the above equation for η with these values yields a viscosity of outer core fluid of 8×10^{15} Pa-s.

Discussion

This is another four orders of magnitude beyond the top end of the 14-orders-of-magnitude range mentioned in the quote above.

That the result of this direct, Noah's-Flood viscosity measurement technique should be higher than the high end of the previously measured range is not surprising. Once it is understood that the true value lies at the high end of the range, then it becomes clear that the viscosity of outer core fluid is strongly pressure dependent. The viscosity of molten iron (believed to be the major constituent of outer core fluid) at 2000 K and atmospheric pressure (1×10^5 Pa) is 6×10^{-3} Pa-s. Core temperatures are a factor of two above this, which will reduce the viscosity. So the much larger, measured high-end viscosity (i.e., 8.6×10^{11} Pa-s) must result from the high pressures characteristic of the outer core (say 2×10^{11} Pa). The viscosity of molten iron thus appears to increase by something like 14 orders of magnitude when the pressure is increased by roughly 6 orders of magnitude. Since pressures in the core would necessarily have been higher than normal during the Flood, as discussed in the previous chapter, it is to be expected that the viscosity found from Noah's observations will be higher than any value measured in modern times.

A Congealed Outer Core

Once it is realized that the viscosity of the outer core is strongly pressure dependent, a second means of holding the inner core pinned to the mantle presents itself. It becomes possible that the inner core was held in place against the mantle by congealed outer core fluid.

Pressures would have been increasing in the core throughout the compression phase of the gravitational collapse mechanism. This means that outer core viscosities would have been strongly increasing throughout the compression phase. It thus becomes possible for viscosities to have become temporarily so large as to prohibit further macroscopic motion in the core on the months-long timescale pertinent to the Flood. In that case, all currents would have diminished nearly to zero and the inner core would have been suspended in place against the mantle until viscosities had reduced sufficiently during the decompression phase to enable it to begin to fall back toward center.

This possibility implies that the viscosity estimated using the terminal velocity equation above may be significantly too low. That calculation used the travel time from center to the mantle (40 days) to estimate the needed terminal velocity. To say that the outer core congealed, effectively immobilizing the inner core on the timescale of interest to the Flood, is just another way of saying that the terminal velocity had become nearly zero. And in that case, the calculated corresponding viscosity becomes very much larger.

This does not upset either the foregoing analysis or the working hypothesis for inner core displacement on which it rests. Rather, it extends it into the very high viscosity regime. The important point is merely that the 8×10^{15} Pa-s viscosity calculated above should be regarded as a minimum value.

Conclusion

Noah's observations of the Flood provide what is probably the most direct means possible of measuring outer core viscosity. These observations say that this viscosity is very large and strongly dependent on pressure. A minimum order-of-magnitude value for the viscosity under the enhanced core pressures present during the Flood has been found to be 10^{16} Pa-s.

A large outer core viscosity has implications for the origin and sustenance of earth's magnetic field. This topic will be discussed in Chapter 40.

CHAPTER THIRTY-TWO

How to Nudge the Inner Core Off Center

It is now clear how to cause a hemispherical Flood—displace earth's inner core from its usual location at the center of the earth out to contact with the mantle. And we now have a working hypothesis for how such a stupendous feat might be accomplished—via gravitational collapse of the earth following slight displacement of the inner core from center. Remaining to be explained is how the inner core could ever have been nudged off center in the first place.

For gravitational collapse to begin to operate, the inner core must somehow first be moved slightly off center. As long as the inner core remains centered, as it is today for example, gravity squeezes down on it symmetrically from all sides. There is then no net force in any direction, so the inner core just remains on center. For gravitational contraction to be initiated—to trigger the Flood—some asymmetric force is required to push or pull the inner core off center.

How to Push or Pull on the Inner Core

How might the inner core have been moved off center at the time of the Flood? What natural physical force—what natural push or pull—exists which might be applied to the inner core to move it?

There appears to be but one possibility. The only thing which is normally in physical contact with the inner core is the outer core fluid which surrounds it. If this fluid starts to move—to flow—then it has potential to drag the inner core along with it.

Chapter 30 advanced a working hypothesis for how an existing core-wide circulation might be amplified due to gravitational collapse. The question now is how core-wide circulation might be initiated. It turns out that this can be accomplished by pulling or pushing on the solid crust and mantle of the earth.

Forces

Gravity

One way to pull on the crust and mantle of the earth is gravity. This is what the sun is doing to the earth all the time, which is why earth orbits the sun rather than voyaging off through space.

But gravitational attraction is the wrong way to accomplish displacement of earth's inner core. The problem, as mentioned previously, is that gravitational attraction operates on the whole earth, including the inner core and the outer core fluid, so the core moves in concert with the rest of the earth.

Gravitational pull gives rise to no displacement of the inner core relative to the rest of the earth.

Collision and Propulsion

While gravity will not work for the present purpose, there are two ways to push the earth, both of which will, in principle, work. The first is collision of earth with some other object. The second is propulsion of the earth by expulsion of matter, the way rockets are propelled in space. Collision of earth with another space object would imply an impact type of event, like an asteroid impact. Propulsion of the earth would imply a volcano-like event.

The result of both collision and propulsion is to accelerate the solid crust and mantle. A big asteroid, striking earth at the South Pole, will accelerate the solid crust and mantle to the north, as will a big volcano erupting at the South Pole.

Fluid Dynamics

To explore what happens in the core of the earth when the crust and mantle are pushed requires a brief venture into the field of fluid dynamics.

Real-life problems in fluid dynamics are generally not trivial to solve. The problem of the motion of earth's solid inner core in the fluid outer core in response to an acceleration—a push or a pull—applied to the mantle and crust of the earth is, unfortunately, not an exception to this general rule. Numerical methods on a computer are required to solve the partial differential equations which describe the problem. To make headway with the problem in a reasonable allotment of time, I found it necessary to simplify as follows.

Simplifying the Problem

To begin with, I simplified the geometry of the problem by treating the figure of the earth as a perfect sphere. More to the point, I treated the surface of the core-mantle boundary as a perfect sphere. This permitted use of spherical coordinates, greatly simplifying mathematical definition of boundaries and boundary conditions.

Next, I chose to neglect earth's rotation. This neglects the Coriolis and centrifugal fictitious forces. These should be unimportant for the present problem which is only concerned with the initial motion of the inner core from its usual position on center (i.e., at the origin). This simplification allowed the problem to be tackled in two dimensions rather than three.

Third, I replaced the solid inner core with outer core fluid, enabling the entire core to be treated as a homogeneous fluid. I set the initial density of the fluid everywhere equal to the total core mass divided by the total core volume. This simplification allowed fluid motion to be explored apart from gravitational effects.

Even though I initialized the density of the fluid to the average core density everywhere, I went on to solve the compressible fluid problem rather than the incompressible fluid problem. This approach was attractive because a simple equation of state for the fluid could be easily derived from Preliminary Reference Earth Model data.[1] The motion of the fluid

[1]Frank D. Stacey, *Physics of the Earth* (Australia: Brookfield Press, 1992), 454–455.

should be insensitive to whether the fluid is treated as compressible or incompressible according to dimensionless variable analysis. The Mach number in this instance is expected to be less than 10^{-3}.

Setting Up the Simplified Problem

The earth was pushed northward by a force applied at the South Pole in my numerical simulation. Thus the earth was propelled to the north (upward).

At time zero (the start of the Flood), the acceleration was turned on.

Keeping things as simple as possible, the acceleration was assumed to be constant with respect to time, yielding a constant acceleration of earth's mantle and crust.

The acceleration of the mantle-crust was, using Einstein's equivalence principle, treated as a reverse acceleration (a g field) acting on the core fluid with the mantle and crust stationary.

Finite difference (Navier-Stokes) equations were derived and implemented on a 10 by 21 polar grid in Fortran on a Windows DOS Pentium IV platform using gfortran via MinGW32 (specifically, tdm-gcc-4.7.1-2).

Viscosity and Acceleration

I chose a viscosity of 1×10^{10} Pa-s for the core fluid, and an acceleration of 1×10^{-10} m/s^2 for the mantle/crust. These specific values were chosen only to keep the calculation within the realm of the physically possible, not to correspond to actual values at the time of the Flood. The actual values of the viscosity and the acceleration corresponding to the Flood are not known at present. The specific values used in the computer calculation are relatively unimportant at this stage. Their effect is to control the rapidity with which motion of the core fluid develops. At present, it is only the form of the motion which is of interest.

Result

Figure 32.1 shows the result of my computer calculation following ten days of constant acceleration at 1×10^{-10} m/s^2. A core-wide, toroidal circulation results, exactly as needed. But the direction of the flow is a little surprising. Notice, again, that the applied acceleration is upward in the figure; the earth is being pushed upward at the South Pole. Thus

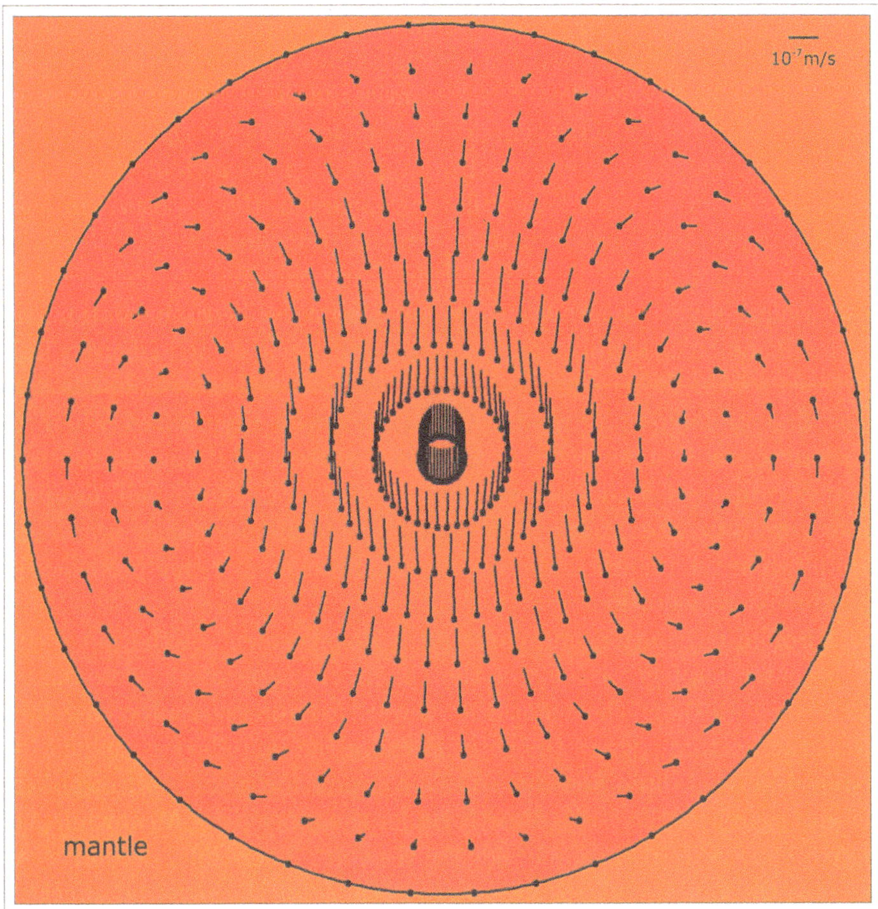

Figure 32.1: Computer simulation of outer core fluid motion resulting from acceleration of earth's mantle toward the north (i.e., upward) at 1×10^{-10} m/s^2 for ten days. The toroidal flow apparent here is pictured in a stationary earth reference frame. Dots mark the grid points used in the finite difference calculation. A 10 by 21 grid space was used. Lines extending from the grid points show the directions and relative magnitudes of velocities of the fluid. The scale bar corresponds to a fluid speed of 1×10^{-7} m/s.

earth as a whole is gaining an upward velocity (not seen here because earth is being viewed in a stationary earth reference frame). But, somewhat unexpectedly, the fluid at the center of the earth is moving upward at an even greater velocity than the mantle and crust.

This immediately shows that the inner core can indeed be displaced from center by acceleration of the mantle and crust. But, according to the numerical simulation, it will not be displaced toward the source of the acceleration as might be expected from naive considerations of the inner core's inertia alone. Rather it will be propelled in the opposite direction even more rapidly than the mantle is being propelled.

This motion of the fluid is a transient behavior. It will not persist indefinitely. In steady state, for a constant applied acceleration, the fluid will obviously simply stratify horizontally, increasing in density from north to south. But to get to that steady state configuration, the fluid must first undergo the toroidal motion shown in Figure 32.1 when the acceleration is first applied. And in the process of doing so, the concentrated upward fluid velocity at the center of the torus will drag the inner core off center toward the opposite side of the mantle even more rapidly than the mantle and crust are simultaneously being propelled by the applied acceleration.

Toroidal circulation and consequent displacement of the inner core from center happen because of the spherical shape of the core-mantle boundary. Pushing the mantle northward compresses outer core fluid along the bottom edge of the core. Because the boundary is shaped like a spherical bowl, fluid flow due to this compression is focused toward the symmetry axis of the bowl, directly under the inner core. Meanwhile, the opposite happens at the upper edge of the core. There fluid decompresses because the mantle is moving away. Because of the inverted spherical bowl shape of the upper core-mantle boundary, fluid above the inner core diverges into the low pressure zone above it. The net result is toroidal circulation of outer core fluid.

Conclusion

Fluid dynamics calculations of outer core fluid in a spherical cavity the size of earth's core show how the inner core may have been moved from its usual central position at the time of the Flood. It may have been displaced by concentrated rising currents in the center of a fluid torus incited by an acceleration of the mantle and crust of the earth due to an impact or volcanic event at the surface of the earth.

Such toroidal circulation would have been ideally suited to subsequent amplification in a consequent gravitational collapse as described previously. And amplification is essential. While a small displacement of the inner core from its usual central position may reasonably be ascribed to a push applied to the crust and mantle of the earth, displacement of the inner core from center all the way to the mantle cannot. The problem is that the push gives most of its energy to acceleration of the whole earth and relatively little of its energy to relative outer core fluid motion.

To get a feel for the kinetic energy acquired by the outer core fluid due to acceleration of the earth, I calculated the kinetic energy of relative core fluid motion versus time for the case presented in Figure 32.1 (page 249). After ten days of acceleration at 1×10^{-10} m/s^2, the kinetic energy of the whole earth was 2.2×10^{16} joules, while the kinetic energy of the fluid motion relative to the mantle was only 1.4×10^9 joules. That is, only about sixty-four billionths of the acceleration energy had gone into relative fluid motion.

To get the inner core to travel to the mantle merely by pushing on the earth, the core fluid would need to be given an energy of at least 7.4×10^{28} joules, as will be shown in the next chapter. This is about one hundred times more energy than is needed to boil earth's oceans completely away. But to give the outer core fluid this much energy so it can carry the inner core to the mantle simply by pushing on the earth requires, according to my computer fluid dynamics program, input of another sixteen million times as much energy. This much energy applied to the earth during the Flood would surely have rendered the planet uninhabitable.

It does not help to try to spread the needed energy application out over a longer interval by maintaining a more modest acceleration for a very long time. The toroid which moves the inner core off center is only a transient, as just discussed. The steady state behavior of the fluid under a constant applied acceleration is simply density stratification with no large-scale motion. Thus initial toroidal motion necessarily devolves into motionless horizontal stratification. The kind of large-scale motion of the fluid required to move the inner core off center will be found only near the beginning of an applied acceleration.

While the inner core may be nudged off center by toroidal fluid motion produced by pushing on the earth, it cannot be moved all the way to the mantle simply by pushing on the earth. Pushing on the earth serves only as the trigger mechanism for ultimate displacement of the inner core to the mantle via gravitational collapse of the earth.

Before closing this chapter, notice that in addition to explaining how the inner core may be moved from center, the fluid dynamics calculation presented in this chapter shows that the inner core will impact the mantle on the opposite side of the earth from the applied acceleration. This feature is of large significance to validation of this inner core displacement mechanism, and ultimately to validation of the overall hemispherical Flood model, as will be shown in Chapter 34.

CHAPTER THIRTY-THREE

CHECK: ENERGY CONSIDERATIONS

A first, simple check of the gravitational collapse mechanism for moving the inner core to the mantle may be performed by asking whether the energy produced by gravitational collapse is sufficient to supply the energy needed to raise the inner core to the mantle. A complete balance of energy cannot be calculated at this stage because a number of items, such as the ultimate compression of the core, are not yet known. So the check merely asks whether the energy released by gravitational collapse seems likely to be in the same ballpark as the energy needed to move the inner core to the mantle.

Energy Needed

A ballpark figure for the energy needed can be obtained by calculating the work done against gravity in moving the inner core from center out to 2250 kilometers. I performed this calculation numerically, using a spreadsheet, via the equation:

$$W_g = \int_0^{2250km} (F_g - F_b)\, dr. \qquad (33.1)$$

F_g is the force due to gravity, and F_b is the force of buoyancy experienced by the inner core in the outer core fluid. I used the density figures for the inner and outer core and whatever other parameters were needed from the Preliminary Reference Earth Model (PREM) in this calculation.[1] I found 7.4×10^{28} joules.

How much energy is this?

This is the energy released by detonation of about a trillion 15-megaton fusion bombs. It is enough energy to supply the United States energy demand, at current rates of consumption, for more than five billion years.

Clearly, a very substantial energy source is needed to fuel this process.

Energy Supplied

A ballpark figure for the energy supplied by gravitational collapse can be obtained by calculating the shrinkage of earth's radius due to simply swapping the (expanded) inner core with an equal size fluid sphere from out in contact with the mantle, leaving all other parameters (such as the core pressure profile) unchanged. This will seriously underestimate available energy, of course. Core pressures would necessarily have increased substantially during the Flood as discussed previously. But it should still serve to provide the ballpark figure being sought.

I wrote a computer program to carry out this calculation. It sectioned the inner core into cubes 50 kilometers on a side. These cubes were individually moved to the inner core's new location in contact with the mantle in 10 meter steps. The volumes of the cubes were recalculated at each step making use of PREM incompressibility data. Once an inner core cube had come to its final position, an equal volume cube of outer core fluid was moved in 10 meter steps back to the starting position of the inner core cube, the new compressed volume being calculated at each step. The difference between the initial inner core cube volumes and the final fluid cube volumes gave the vacated volume needing to be filled by shrinkage of earth's radius. This turned out to be 5.73×10^{16} cubic meters. Filling this volume was found to release 3.2×10^{28} joules of gravitational potential energy. (Notice, in passing, that the mass needed to fill this [minimum] vacant volume was 7.0×10^{20} kilograms. This corresponds to shrinkage of earth's radius by just 530 meters.)

[1]Frank D. Stacey, *Physics of the Earth* (Australia: Brookfield Press, 1992), 454–455.

Conclusion

A ballpark figure for the energy needed to move the inner core against gravity to the mantle has been found to be 7.4×10^{28} joules. A ballpark figure for the energy supplied by gravitational collapse by merely swapping the inner core with an equal volume of outer core fluid from out near the mantle has been found to be 3.2×10^{28} joules. These two huge energies differ by just a factor of two. They are clearly in the same ballpark.

Thus it checks. The proposed gravitational contraction mechanism appears to be energetically matched to the inner core displacement mechanism.

Chapter Thirty-Four

Check: Flood Antipodes

The idea that the inner core was displaced from center by an impact or volcanic process pushing on earth and resulting in the inner core subsequently colliding with the mantle on the more or less exact opposite side of the globe at the time of the Flood leads to what is probably the most exacting check of the overall hemispherical Flood model.

The fact is that real historical events tend to leave a trail, and the trail can often be discerned long after the event. In the present case, volcanic processes tend to alter the surface of the earth, producing mountains and calderas for example, while asteroid impacts tend to leave impact craters. Thus some such surface feature should be found corresponding to the push which triggered the Flood. And on the more or less exact opposite side of the earth should be found clear evidence of extensive volcanism at the surface, above where the inner core impacted the mantle.

Extensive volcanism is expected because the collision between the inner core and the mantle is expected to have been energetic. At the end of its trajectory, just prior to collision with the mantle, the inner core would have had some velocity, v_f, and an associated kinetic energy given by:

$$E = \frac{M {v_f}^2}{2}. \tag{34.1}$$

The mass of the inner core, M, is 9.8×10^{22} kg. The final velocity, v_f, is not known, but the average velocity previously calculated (page 243) should put us in the right ballpark. The result is 2.1×10^{22} joules.

This is the sort of energy which results from impact of a large (1 km or more) asteroid at the surface of the earth. Taking 2×10^5 megatons to be the threshold of global disaster for asteroid impacts, it is the energy equivalent of twenty-five global disasters. It is what one gets from detonation of five million 1-megaton nuclear bombs.

Extensive volcanism is expected for such an energetic collision between the inner core and the mantle because the dissipation of inner core kinetic energy during the collision will likely result in upward flexure and fracture of the overlying mantle and crust. Considerable heat will be generated in the flexed and fractured rock, and as heat travels slowly in the earth, onset of long-sustained (millennia-scale) volcanism at the surface seems likely. Thus the proposed Flood mechanism predicts reasonably well-characterized surface features in a unique antipodal relationship (Figure 34.1).

Because flooding seems to have been confined more or less to the Northern Hemisphere during the Flood, the inner core is expected to have risen roughly along earth's rotational axis into the northern hemisphere of the outer core. Thus, the evidence of extensive volcanism is likely to be located in or near the Arctic region of the globe. And this implies that the source of acceleration should be expected on the opposite side of the earth, in or near the Antarctic region of the globe.

But it is possible to set much more stringent limits than this on where the site of extensive volcanism can be located. I will show shortly that the idea that the waters of the Flood were raised by displacement of earth's inner core stipulates that this site must be located in a narrow band, a constant angular distance from the landing place of the ark.

In this way, the proposed Flood mechanism furnishes a rigorous test of its own validity. It predicts existence of an antipodal pair of high-energy surface features within narrowly delineated geographical coordinates. The check this time is whether suitable antipodal surface features can be found within the allowed angular distance.

The Ark's Resting Place

To carry out this check, it is important to be as accurate as possible about the location of the resting place of the ark, both as a geographical reference point, and as a means of estimating the most probable actual depth of water at that point. (Recall that merely an estimated depth of water for the entire Ararat region has been used in the development of this mechanism for the Flood so far.)

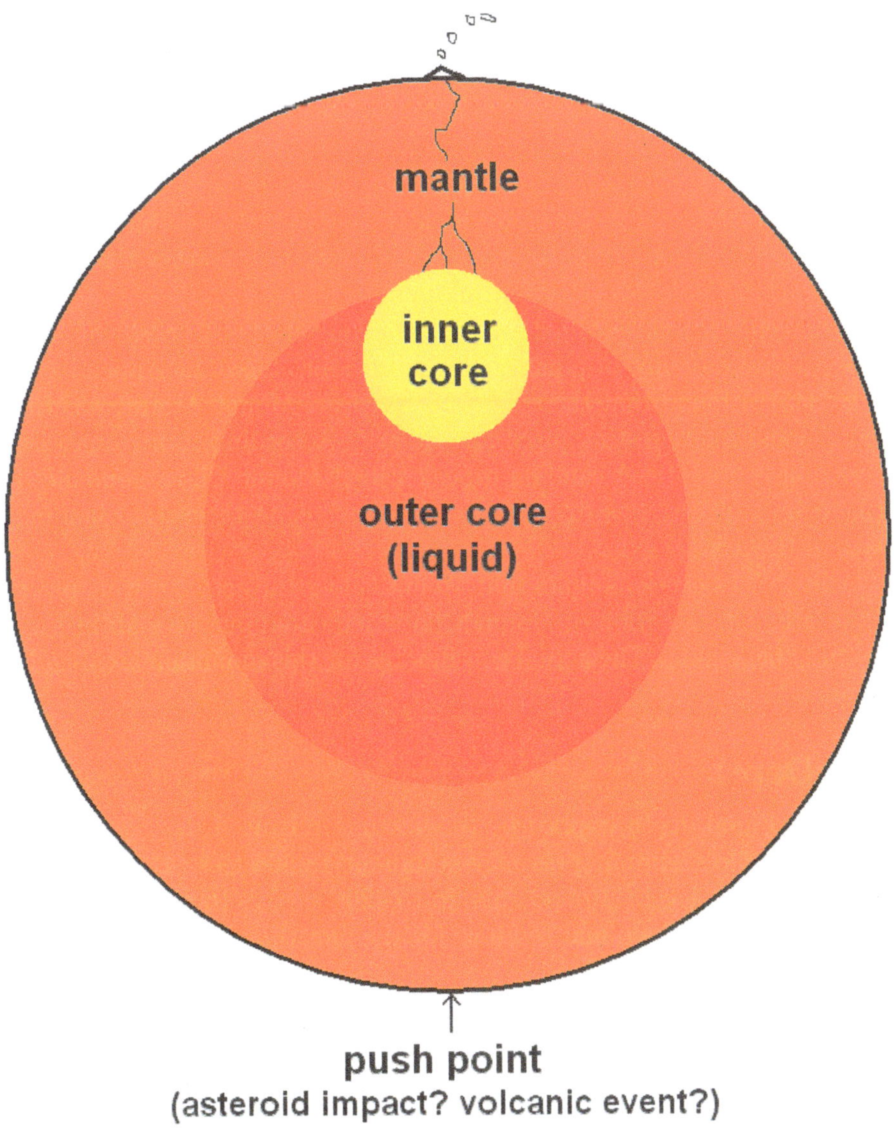

Figure 34.1: Conceptual diagram illustrating the expectation of antipodal surface features. Not to scale.

I tackled this problem some years ago.[1] I used a number of clues regarding the mountain the ark landed on given by the Genesis narrative. For example, the text informs us that neighboring mountains first became visible 72 days after the ark had grounded on its mountain, during which time "the water decreased steadily" (Genesis 8:5). This immediately implies that the ark grounded upon the tallest mountain in its general vicinity.

As another example, Genesis 8:4 says that "the ark rested in the mountains of Ararat." This makes it clear that the ark must have landed on a mountain which was in among other mountains rather than on a mountain which was situated out on a plain.

I ranked 1441 peaks in the Ararat region relative to a set of such clues. Details of my original analysis are reproduced in Appendix F (page 345). I found that the traditional favorite, Mount Ararat, was the second most likely resting place for the ark. According to my analysis, a previously unnoticed candidate peak called Mount Cilo (pronounced JEE-lo in Turkish) was 62 times more likely to be the actual landing place of the ark.

The present check assumes that Mount Cilo is the actual resting place of the ark. The location of Mount Cilo is indicated on the map in Figure 28.1 (page 221). The actual coordinates for this mountain are 37.5 N and 44.0 E.

The Depth

Next, the Flood depth implied by this mountain is needed.

Mount Cilo's summit presently stands at 13,566 feet (4135 meters) above mean sea level.[2] Round this up to 4200 meters to allow for some erosion since the Flood and to allow for the fact that the mountain was initially covered by the Flood.

The Check

Figure 34.2 shows the gravitational potential calculation once again. The horizontal dashed line marks the Flood depth at Mount Cilo. The vertical dotted line marks the location of Mount Cilo relative to the North Pole.

[1]Gerald E. Aardsma, "The Ark on Ararat?" *The Biblical Chronologist* 3.2 (March/April 1997): 1–12. www.BiblicalChronologist.org.

[2]Map TPC G-4B, Defense Mapping Agency Aerospace Center, St. Louis, Missouri.

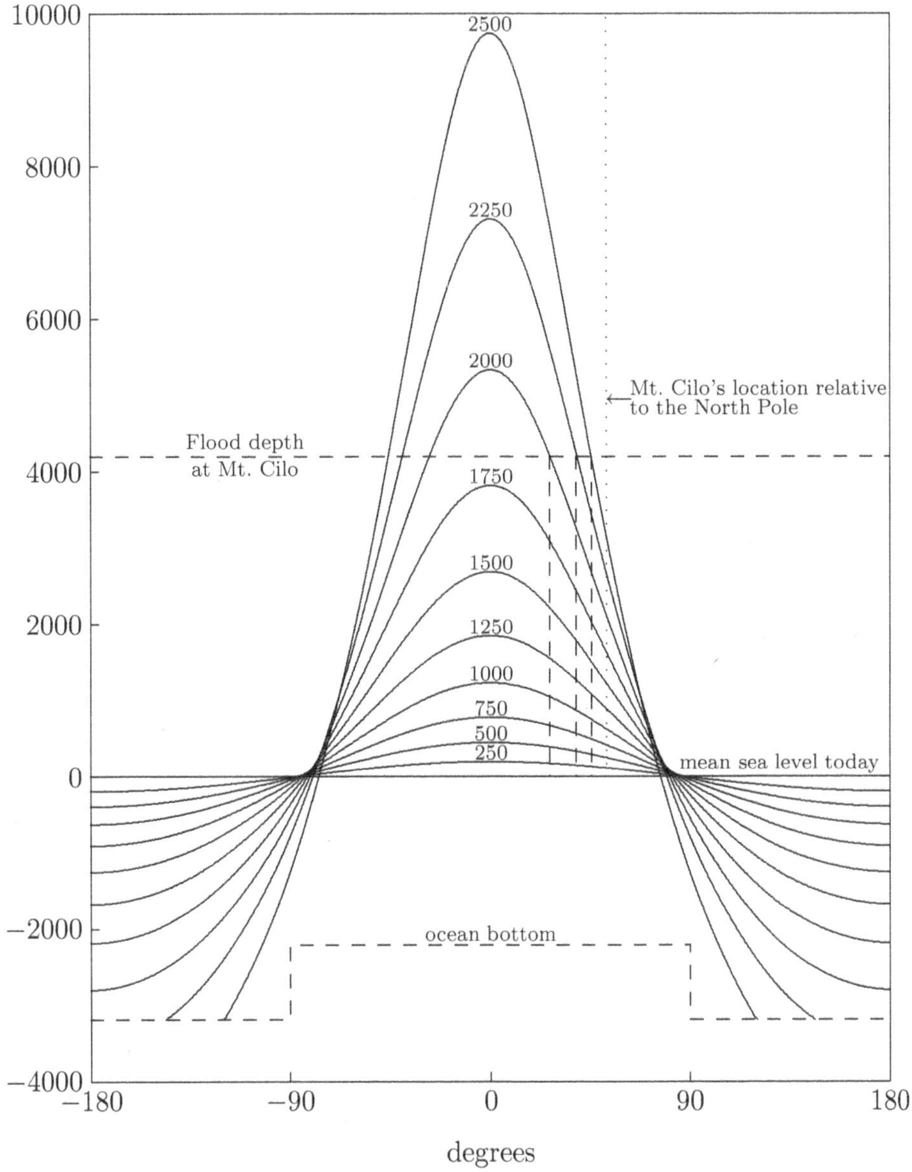

Figure 34.2: Depth of the water of the Flood (in meters above mean sea level) for various displacements of the inner core (in kilometers).

Notice that while a 1800 kilometer displacement of the inner core produces a Flood which is deep enough at the North Pole, it does not produce a Flood which is deep enough at Mount Cilo. In fact, none of the depth curves is deep enough at Mount Cilo. What this is saying is that the inner core did not rise to the mantle directly beneath the North Pole.

In the construction of Figure 34.2, I treated the displaced inner core as if it had risen directly beneath the North Pole for two reasons. First, because it is clear that it must have risen roughly along earth's rotational axis toward the North Pole to explain the Flood's extension to such widely separated points in the Northern Hemisphere as Minnesota and Mesopotamia while at the same time explaining its extension into the Arctic region. And second, because it was already hard enough to communicate the basic physics involved without having to introduce the added complication of a latitudinal offset. But now it is time to be more rigorous.

The earth was pushed upon wherever it was pushed upon. The inner core would have risen opposite this push point. Thus the water of the Flood would have been deepest opposite the push point, not over the North Pole.

This can be taken into account by simply interpreting the "degrees" on the horizontal axis of Figure 34.2 as angular distance from the displaced inner core, rather than as angular distance from the North Pole. Strictly speaking, this is not quite proper because the elevated ocean floor must remain centered on the North Pole rather than moving with the displaced inner core. But the angular distance of the displaced inner core from the North Pole seems likely to be small enough to ignore this at the level of accuracy afforded by this simple model.

Adopting this new definition for the horizontal axis allows the maximum angular distance from the displaced inner core to Mount Cilo for various displacements of the inner core to be read off from the figure. These angular distances correspond to the horizontal axis coordinate (the degrees) at which a given water depth line intersects the horizontal dashed Flood-depth-at-Cilo line.

Only three of the water depth curves intersect the horizontal dashed line. The intersection points are shown in the figure by the three vertical dashed lines extending down to the horizontal axis.

The water depth curve corresponding to a 2000-kilometer displacement of the inner core intersects the horizontal line at 27 degrees. This means that, if the inner core had been displaced a maximum of 2000 kilometers from center at the time of the Flood, then the displaced inner core could not have been farther from Mount Cilo than 27 degrees to get deep enough water at Mount Cilo. Similarly, the 2250 kilometer displacement curve intersects at 39 degrees, and the 2500 kilometer curve gives 46 degrees.

Extensive Volcanism Site

These angular distances are sketched, for a general northerly direction, on the globe in Figure 34.3. We must now ask whether any regions of extensive volcanism exist within the pie slice of earth's surface defined by the X and the 2500 kilometer arc in this figure.

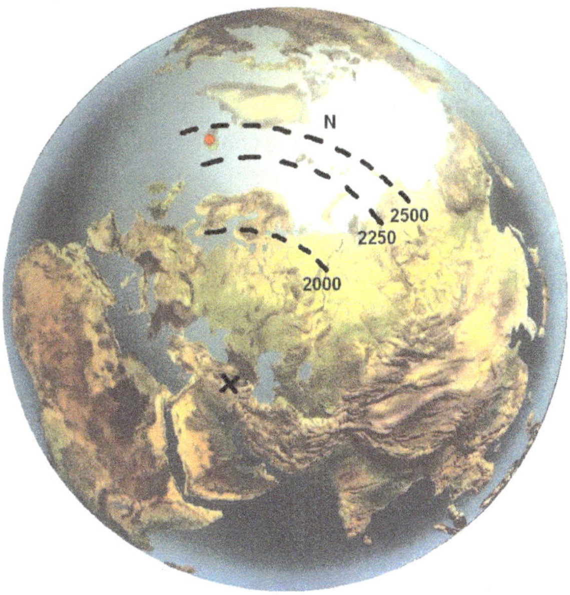

Figure 34.3: View of the earth showing maximum angular distances from Mount Cilo to the extensive volcanism site for three displacements of the inner core (in kilometers). (X = Mount Cilo, N = North Pole, red dot = Iceland; figure adapted from *Reader's Digest Great World Atlas*, 1963.)

I can find but one candidate. This is Iceland, shown in the figure by a red dot.

Iceland is one of the most volcanically active spots on earth today:[3]

> More than a hundred volcanoes, some still active, make Iceland one of the most volcanic regions of the world... The volcanic rocks heat countless hot springs and geysers... The warm waters are piped to heat buildings and hothouses, in which vegetables, fruits, and flowers grow the year around. In some places water is piped through the soil to warm it for growing green crops.

Despite much effort, Iceland's anomalously high volcanic activity has remained without cogent explanation to the present time. Iceland sits on a continuation of the Mid-Atlantic Ridge into northern latitudes. The ridge is known as Reykjanes Ridge to the south of Iceland, and Kolbeinsey Ridge to the north. This line of deep ocean ridges marks the boundary between adjoining slabs of rocky crust. The slabs, or plates, are like the pieces of the shell of an egg which has been cracked. The plates of the earth grind and push on each other at these boundaries, making the boundaries generally more active than the rest of the surface of the earth in regard to earthquakes and volcanoes. I think it is common to attempt to explain Iceland's very high volcanic activity as due to its location on this boundary between plates. This is not very satisfactory, however. There are many plate boundaries girdling the earth. What makes Iceland so uniquely volcanically active?

It seems clear that Iceland is the extensive volcanism site which the Flood model predicted should be found, and that collision of the inner core with the mantle beneath Iceland just 5500 years ago is the cause of its currently observed anomalously high volcanic activity.

Push Point Site

Iceland lies between longitudes 13 and 25 W, and latitudes 63 and 67 N. Overlapping the corresponding area on the opposite side of the globe is the Balleny hotspot with its volcanic islands (Figure 34.4). Thus the Balleny Islands fulfill the model's expectation of an antipodal volcanic or impact surface feature. Evidently the Flood was triggered by a volcanic event, not an impact event.

[3] "Iceland," *Compton's Encyclopedia*, vol. 11 (Chicago: William Benton, 1972) 10.

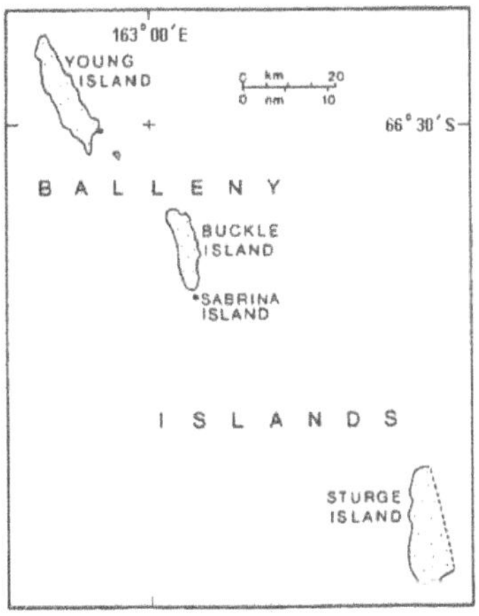

Figure 34.4: Balleny Islands (http://en.wikipedia.org/wiki/Balleny_Islands).

Conclusion

The angular distance between Iceland and Mount Cilo corresponds to a 2400 kilometer displacement of the inner core. This is significantly beyond the 2258 kilometer boundary between no contact and collision of the inner core with the mantle. Thus Iceland also fulfills the proposed Flood mechanism's expectation that the inner core did, in fact, collide with the mantle at the time of the Flood. This remains true even if the landing place of the ark is switched from Mount Cilo to Mount Ararat.

It is now possible to estimate, based on the angular distance between Iceland and Mount Cilo, that the collision embedded the upper 150 kilometers of the inner core in the mantle.[4] The collision was, therefore, an energetic one, consistent with the volcanic expression of excess energy observed for Iceland today, and consistent with Noah's observation that the Flood persisted at full depth for 110 days.

Finally, the check also fulfills expectations of high outer core viscosity. Because the earth is rotating, Coriolis and centrifugal forces arise, bending trajectories as observed in the stationary earth reference frame. Thus, these fictitious forces have potential to destroy the antipodal relationship between the push point and the extensive volcanism site. High viscosity prevents this by damping motion due to these forces.

[4]Geophysicists will be interested to note that this corresponds to penetration through most of the D" layer.

To explore the dependence on viscosity of the antipodal relationship between the push point and the extensive volcanism site, I wrote a computer program to calculate the trajectory of the inner core for an assumed constant total radial force (applied plus gravitational) propelling the inner core toward the mantle for an assumed constant viscosity. Coriolis, centrifugal, and viscous drag forces were calculated and applied stepwise along the trajectory. Outer core fluid was treated as stationary since the impact of drag was of interest only for non-radial motion of the inner core. The constant radial force was adjusted to achieve a 40 day travel time for the inner core from center to impact with the mantle for various choices of viscosity. Terminal velocity was rapidly achieved in all cases of interest, resulting in a constant inner core velocity for most of the trajectory.

I found that, while the antipodal relationship was lost for viscosities much below 5×10^{14} Pa-s, it was well preserved at 10^{16} Pa-s, the minimum order-of-magnitude viscosity of outer core fluid during the Flood previously calculated (page 244).

Thus this exacting check of this overall proposed mechanism of the hemispherical Flood model seems satisfied in every way.

CHAPTER THIRTY-FIVE

CHECK: ZOOGEOGRAPHY AND NOAH'S FLOOD

Once it is known that the inner core rose beneath Iceland, it becomes possible to check the hemispherical Flood model in another way, this time from the field of biology.

The effect of the Flood on species diversity should still be imprinted on the planet. Specifically, the Northern Hemisphere should have lost many species as a result of the depth and persistence of the Flood there, while species in the Southern Hemisphere should have gotten off much better. More precisely stated, the region of the globe which was deeply flooded, centered on Iceland, should show low species diversity today compared to the rest of the globe.

Zoogeography

Zoogeography is the science of mapping out regions having distinct faunas. The modern discipline of zoogeography traces its roots back to 1876 with the publiation of Alfred Russel Wallace's two-volume set, *The Geographical Distribution of Animals.*[1] The modern discipline incorporates the advantages of phylogenetic relationships, modern statistical methods, and computers to process all the data. Nonetheless, its results differ little

[1]Sandra Knapp, "What, Where, and When?" *Science* 341 (13 September 2013): 1182–1184.

from Wallace's in broad outline.[2]

Result

Figure 35.1 shows the results from zoogeography with lines of constant angular displacement from Iceland hand sketched by me. While continental topography needs to be included to do this to high precision, the fit of the data to the model is obviously good in panoramic outline.

The most striking observation from zoogeography, whether using Wallace's original map of zoogeographic regions or more technologically sophisticated modern maps of evolutionary uniqueness, is the reduced variety in the north relative to the south. The modern discipline sees this disparity more clearly than ever, and struggles to find a cogent explanation of it:[3]

> ...the subtle differences in the phylogenetic composition of assemblages over the Northern Hemisphere as a whole might be a consequence of a high degree of connectivity and range dynamics. Low rates of extinction resulting from greater climatic stability in the Southern Hemisphere could also have contributed to this pattern by allowing species that belong to ancient clades to persist through time.

The hemispherical Flood Model does not struggle to explain this observation—it predicts it.

Conclusion

Zoogeography bears elegant testimony to a relatively recent hemispherical Flood.

---

[2]Ben G. Holt et al., "An Update on Wallace's Zoogeographic Regions of the World" *Science* 339 (4 January 2013): 74–77.

[3]Ben G. Holt et al., "An Update on Wallace's Zoogeographic Regions of the World" *Science* 339 (4 January 2013): 77.

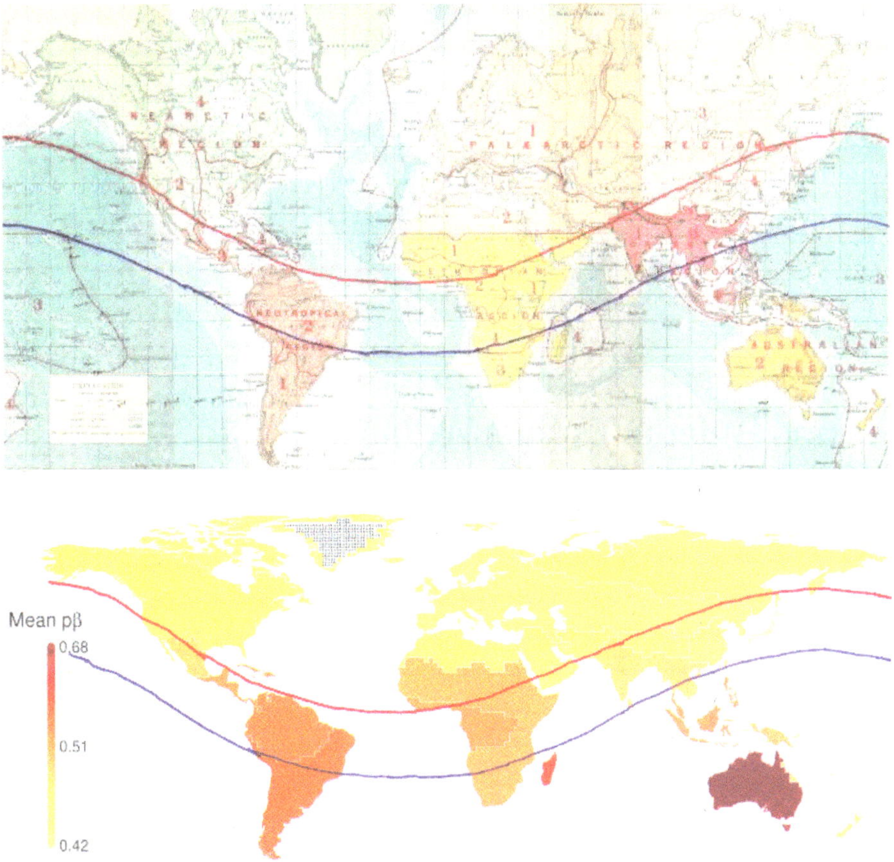

Figure 35.1: **TOP:** Zoogeographic regions defined by Wallace with lines showing the geographical extent of different Flood regimes resulting from the Flood depth curves of Figure 34.2 (page 261) for Iceland, hand-sketched by me. The bottom line is 88 degrees from Iceland. This line flooded briefly to a depth of roughly 70 meters early in the Flood, but spent from Day 40 to Day 150 below -600 meters. The top line is 62 degrees from Iceland. It corresponds to a Flood depth from Day 40 to Day 150 of roughly 1900 meters. As the average elevation of the continents is just 840 meters, survivability is expected to have been low above this line. [Figure adapted from Sandra Knapp, "What, Where, and When?" *Science* 341 (13 September 2013): 1183.] **BOTTOM:** Evolutionary uniqueness [yellow (less) to red (more)] with same Flood lines sketched by me as in the map above. [Figure adapted from Ben G. Holt et al., "An Update on Wallace's Zoogeographic Regions of the World" *Science* 339 (4 January 2013): 76.]

Chapter Thirty-Six

Conclusion to Part 4

An evidently energetic volcanic event occurred at Balleny Islands propelling earth in space 3520 B.C.

Because of the spherical shape of earth's core-mantle boundary, acceleration of the earth due to the Balleny eruption produced a transient toroidal flow of outer core fluid. The concentration of fluid flow near the symmetry axis of the resultant torus displaced the inner core from its usual central position.

Fluid compression in the volume vacated by the inner core transferred momentum to the inner core, driving it farther away from center and making way for the infall and compression of yet more fluid. As the fluid compressed, earth contracted, releasing some of the enormous energy stored in its gravitational field to drive the inner core toward the mantle. The inner core ultimately collided with the mantle, embedding itself roughly 150 kilometers in the mantle beneath Iceland.

The gravitational anomaly (i.e., the geoid warp) caused by the off-center inner core produced an ocean-deep mountain of water in the Northern Hemisphere while partially emptying ocean basins in the Southern Hemisphere. This hemispherical mountain of water was the Flood recorded in Genesis.

The Balleny Islands hotspot marks the site of the initial propulsion which triggered the Flood. Iceland marks the site of the extensive volcanism resulting from energetic collision of the inner core with the underside of the mantle. The antipodal relationship of these two sites, together with their location within narrow geographic limits prescribed by Noah's observational data and the known laws of science, plus the otherwise anomalously high, ongoing volcanic activity of Iceland today, corroborate the overall hemispherical Flood model.

Further development of this model promises to teach us much about our planet, especially its inner workings.

PART 5

THE NATURE OF NOAH'S FLOOD

CHAPTER THIRTY-SEVEN

WHAT NOAH'S FLOOD WAS NOT

It is now clear that the Genesis account of Noah's Flood is not a myth. It is also now clear that the Genesis Flood was not a global, earth-shattering, tectonic cataclysm responsible for most of earth's sedimentary rocks and their entombed fossils.

Myth?

The account of the Flood found in Genesis has now been shown to be quantitatively explicable in terms of displacement of the inner core of the earth 3520 B.C. This means that the existence of a solid inner core and a liquid outer core of the earth is embedded in and presupposed by the Flood narrative. Noah's observations cannot be explained apart from these concepts, while once they have been given, it is possible to explain Noah's observations with mathematical precision.

But these features of the interior of the earth have been known to science for less than a century. Thus it seems that the narrative of the Flood which Genesis preserves cannot be anything other than an eyewitness account.

The argument yielding this conclusion is actually far stronger than that which has just been presented above. The Genesis Flood narrative has now been found to be scientifically explicable, yielding testable predictions unforeseen by the narrative (e.g., antipodal surface features within narrow geographic coordinates), which succeed brilliantly.

To demonstrate this predictive success, not only must the existence of the inner and outer cores of the earth be known, but also their diameters must be known, and their densities must be known, and the fact that the earth is a sphere must be known, and the equation describing the force of gravity must be known, and Newton's Laws of motion must be known, and the volume of water in the oceans of the world must be known, and the area of the continents must be known, and their relative distribution in the Northern and Southern Hemispheres must be known, and their mean height above sea level must be known, and the solutions of fluid dynamics equations requiring a computer to calculate must be known, and...

Did the writer of Genesis know all these things?

Global Earth-Shattering Cataclysm?

The account of the Flood found in Genesis has now been shown to be quantitatively explicable in terms of displacement of the inner core of the earth 3520 B.C. This means that the extent of the Flood was hemispherical, not global. It also means that, painting with a broad brush, it would be far more accurate to describe the waters of the Flood as tranquil than it would be to describe them as cataclysmic. The intact pre-Flood archaeological strata in Mesopotamia, Palestine and elsewhere demonstrate this, as do the pre-Flood stone walls at Céide Fields. The water from the southern oceans flowed up into the Northern Hemisphere at the start of the Flood. The combined oceans stayed heaped up there for months. Then the inner core slowly returned to its normal position on center and the excess water flowed back to the southern ocean basins where it had come from.

The global cataclysmic Flood model of Whitcomb and Morris,[1] the cornerstone of Young Earth Creationism today, is seen to be seriously in error. Genesis does not teach the idea that Noah's Flood reworked the surface of the earth to great depth, producing lots of sedimentary strata and lots of fossils. This notion is extra-biblical. Genesis teaches merely that Noah's Flood was a *flood*. The Genesis Flood narrative talks

[1]John C. Whitcomb, Jr. and Henry M. Morris, *The Genesis Flood* (Philadelphia: The Presbyterian and Reformed Publishing Company, 1961).

neither about tearing up the surface of the earth nor about laying down fossil-bearing strata. It talks only about deep, persistent water. You can read it for yourself in Genesis 6–8.

Conclusion

Noah's Flood is not a myth. The monstrous flood described in Genesis really happened. A hemispherical Flood, occasioned by displacement of earth's inner core to the mantle, was unleashed upon the earth mid-Holocene, 5500 years ago, 3520 B.C.

While the Flood was a catastrophe of global proportions, flooding was roughly hemispherical in extent, not global. The Flood did not rip up the surface of the earth to great depth and redeposit it as the geologic column.

CHAPTER THIRTY-EIGHT

THE TRUE NATURE OF NOAH'S FLOOD

The historical Flood which Genesis records breaks naturally into three stages: waxing, maintaining, and waning (Figure 38.1). These three

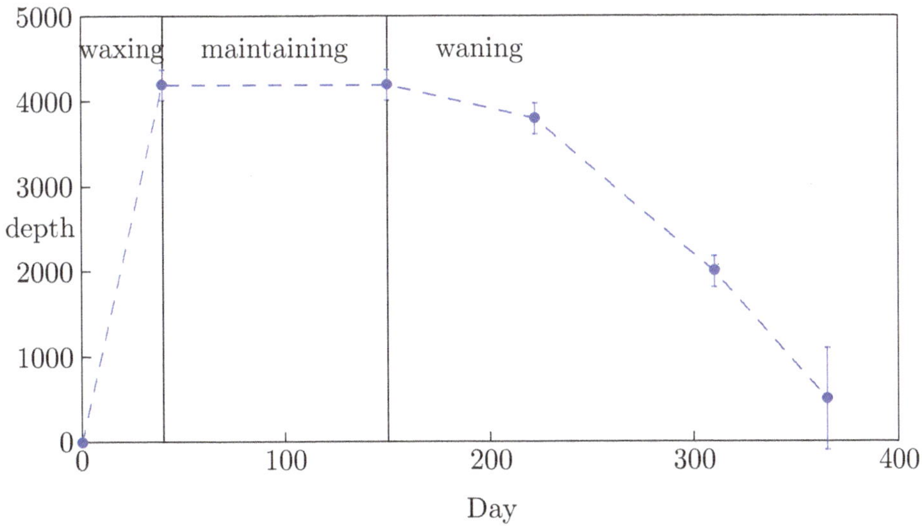

Figure 38.1: The three stages of the Flood illustrated by the depth of the water of the Flood in the vicinity of Mount Cilo (in meters above mean sea level today) versus time. See Appendix G for how the data points are obtained from Genesis. Error bars are $\pm 3\sigma$. The dashed blue lines merely connect the data points.

stages are discussed briefly below, beginning with the middle stage.

Maintaining: The Static Flood Profile

The main picture which should come to mind when thinking about the Flood is the static Flood depth profile which dominated earth from Day 40 to Day 150 of the Flood. In (very rough) first approximation, we may think of the Northern Hemisphere as having been under an ocean of water, while the ocean basins of the Southern Hemisphere were dry. Near the equator, the depth of water was normal. Going north from the equator, the water progressively deepened until eventually more than two normal ocean depths of water were piled up above normal sea level. Going south from the equator, the depth of water decreased until eventually the ocean floors became exposed. This state persisted for 110 days.

This first-approximation picture needs to be corrected when doing more quantitative work, of course. In particular, the fact that the inner core rose beneath Iceland, not beneath the North Pole, needs to be taken into account, as is shown in Figure 38.2. Thus, for example, the "Flood equator"—near which the level of the water was somewhat normal throughout the Flood—would not have corresponded to the geographical equator, and the "Northern Flood Hemisphere" would have been offset from the geographical Northern Hemisphere.

Waxing: The Start of the Flood

When we think about what the coming of the Flood may have been like, the first thing we should notice is that it did not unfold all in a single day of sudden calamity. Noah's observations of the Flood reveal that it took 40 days for the inner core to reach the mantle, and hence it took 40 days for the Flood to reach full depth, as was discussed in Chapter 29.

The fact that the inner core took 40 days to reach the mantle means that the water of the Flood would have come up like an incoming tide at the beach, not like a great, rushing tidal wave sweeping over the land. We now know that the inner core traveled a total of 2400 kilometers in 40 days. Thus its average speed was 2.5 kilometers per hour (1.6 miles per hour). The Flood reached an ultimate depth of 9 kilometers over Iceland, thus its average rate of rise over Iceland was 9.4 meters per hour (31 feet

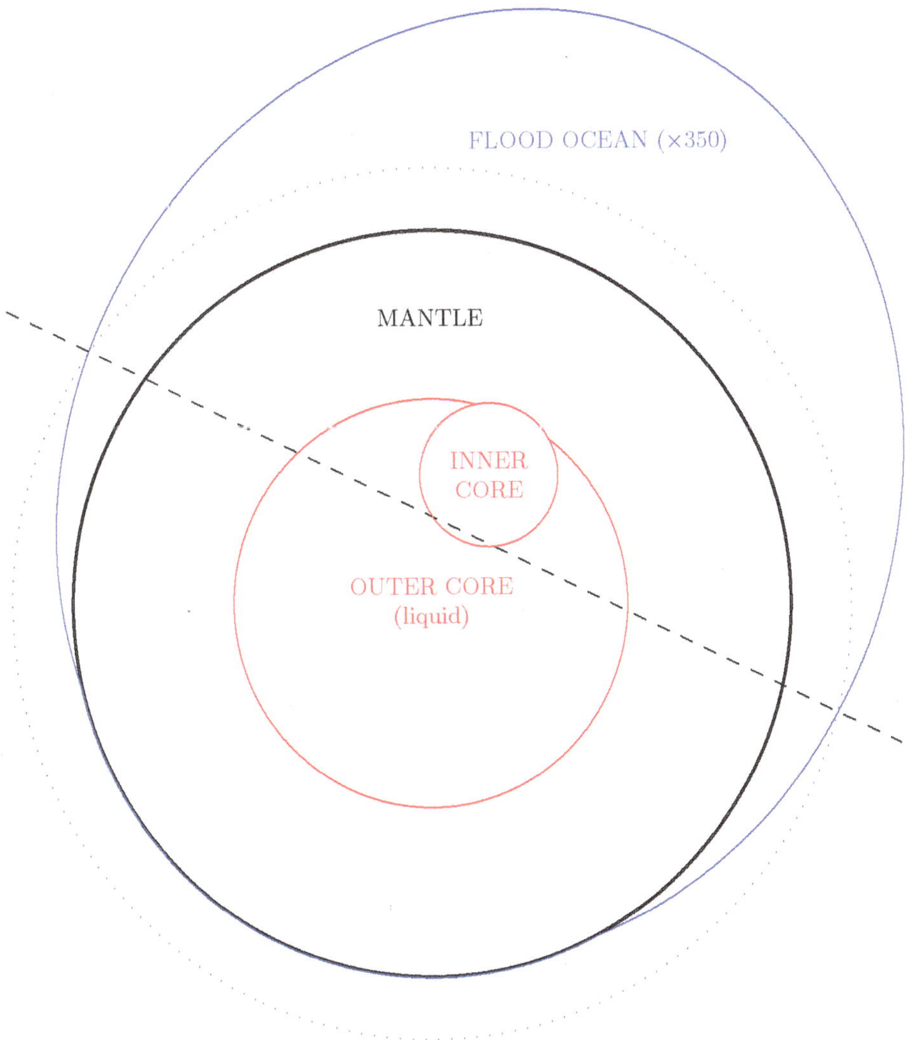

Figure 38.2: Scale cross-section of the earth showing the static Flood depth profile. The Flood ocean depth (blue) has been magnified 350 times relative to scale. The dotted blue line shows the normal ocean depth profile for comparison, also magnified 350 times relative to scale. The north–south polar (rotation) axis is vertical and the equatorial plane is horizontal through the center of the earth. The inner core has risen beneath Iceland. The heavy line is a scale representation of a 30 km thick crust. The plane of the "Flood-equator" (for which ocean depth was normal) is shown by the dashed line. The unequal distribution of continental volume in northern and Southern Hemispheres has been taken into account in the distribution of the ocean water as discussed previously. Notice the dry ocean basin in the Southern Flood Hemisphere.

per hour). Iceland would have experienced the maximum average rate of rise, of course; the Ararat region would have experienced an average rate of rise somewhat less than half this amount, for example.

There would thus have been some opportunity to flee to higher ground. For individuals living near the Flood equator, this strategy would likely have been effective in saving many lives. But for individuals living much to the north, the tide would have just kept coming in higher and higher day after day. Because of the great depth ultimately attained by the Flood in the north, even high mountains would eventually have been covered in many regions (as in the Ararat region), yielding no lasting place of refuge and little chance of survival.

Meanwhile, life would have been subject to a different set of stresses in the Southern Flood Hemisphere. There, thinning of the atmosphere would have resulted in dry, cold conditions and oxygen deprivation for those living far enough south of the Flood equator. In contrast to the mountains of the north, deep valleys, including previously underwater areas such as continental shelves and canyons, would have afforded the best refuge from these conditions.

Waning: The End of the Flood

The waning of the Flood was governed by the return of the inner core to its usual central position. Noah's observations inform us that it took roughly half a year for things to get more or less back to normal at the surface of the earth once the waters had begun to recede.

In the Northern Flood Hemisphere, the drying of the surface of the ground would have been aided by the katabatic wind discussed in Chapter 29.

In the Southern Flood Hemisphere, this same air, humidified by passage over flooded northern realms, would likely have given rise to a prolonged southern hemispherical precipitation event as it was lifted to higher altitudes within the thickening southern atmosphere.

Conclusion

The Noah's Flood event described in Genesis was a mid-Holocene global catastrophe. The physical expression of this catastrophe varied with geographical location and with time throughout the year of the Flood. There was a sharp difference between the Northern Flood Hemisphere and the Southern Flood Hemisphere. The Southern Flood Hemisphere experienced conditions more or less opposite to those experienced in the Northern Flood Hemisphere at any given time.

Flooding covered much of the Northern Flood Hemisphere and was most severe (i.e., several oceans deep) over Iceland. Much of the Southern Flood Hemisphere faced the opposite problem of loss of oceans compounded by loss of atmosphere. The shift of the atmosphere to the north early in the Flood resulted in more than a month of rain in the north and reciprocal drought in the south.

The waxing of the Flood seems properly described as gradual but relentless. The eventual waning of the Flood appears to have been even more gradual, giving rise to dry conditions in the north and wet conditions in the south.

CHAPTER THIRTY-NINE

CONCLUSION TO PART 5

In the absence of a proper scientific understanding of the Noah's Flood event described in Genesis, misconceptions about the Flood have flourished. Most egregious, from a scholarly perspective, has been the near-universal notion of modern science that the Genesis narrative of the Flood is a myth. Surely future generations will shake their heads in wonder that an event so catastrophic and so recent could have escaped detection by so many capable scientists for such a long time—especially in view of the fact that we were blatantly *told* by the ancients that it had happened.

Fortunately, the species as a whole has been wiser than its experts in this particular instance. Millions of lay persons around the globe have persisted in believing the Genesis record of Noah's Flood, sensing that something was amiss with what the experts were telling them. But here, too—much more understandably—misconceptions have been rampant.

A frequently encountered misconception is that the Flood was all over in 40 days. Another is that it was caused entirely by rain. These result, of course, from a failure to give proper attention to the primary account of the Flood found in Genesis.

Another widespread misconception is that Noah's Flood flooded the entire planet. Another is that the only humans in the whole world to survive the Flood were the eight persons aboard the ark. Another is that the only air-breathing, land-dwelling animals in the whole world to survive the Flood were those Noah took aboard the ark. These last three misconceptions flow from the same interpretive error discussed previously. This is the error of failing to recognize that phrases in Genesis which readily appear exhaustively inclusive to us today do not seem to have been understood that way by their original audience. To the original audience, these phrases seem to have conveyed the idea of extraordinary largeness, not exhaustive inclusiveness. Where, for example, we all too easily hear God saying, in Genesis 6:17, "I'm going to wipe the earth clean of every last breathing thing," it appears they heard, "I'm going to bring about a destruction of life on earth the likes of which you have never seen nor can even begin to imagine."

To avoid misconceptions about the Flood and gain a proper understanding of our planet, including the Holocene epoch in which we now live, it is necessary to give proper place to the report of the Flood which is recorded for humankind in Genesis. To avoid misconceptions and gain a proper understanding of the narrative of the Flood in Genesis, it is necessary to give proper place to modern science. These two sources of knowledge relevant to the world in which we live are complementary, not contradictory. To champion either one of them at the expense of the other—whether through ignorance, or arrogance, or prejudice, or ideology, or misplaced zeal, or whatever else—is a sure-fire means of missing out on the truth.

Part 6

The Hazard of Noah's Flood

Chapter Forty

Was Noah's Flood a Singular Event?

Hurricane Katrina, the deadly hurricane which devastated the Gulf Coast region of the United States in August 2005, is recognized to be a specific example of the hurricane class of earth catastrophes. The 1980 eruption of Mount St. Helens is recognized to be a specific example of the volcano class of catastrophes. The San Francisco earthquake of 1906 is recognized to be a specific example of the earthquake class of catastrophes. Was Noah's Flood, the catastrophe which all but extinguished civilization 3520 B.C., a one-of-a-kind event, or is it too just a specific example of a previously unrecognized class of earth catastrophes?

Before proceeding to tackle this question, a name is needed for this new (hypothetical) class of earth catastrophes. I suggest adoption of the name "Noahic Events" in honor of the archaeo-scientist who first recorded observations of a member of this class.

Following are six independent arguments yielding the conclusion that Noah's Flood is just the most recent example of a Noahic Event.

Argument 1

The mechanism of the Flood which has been presented in this volume seems to imply that Noahic Events can occur at any time.

According to this mechanism, Noahic Events are caused by the inner core being displaced from center by acceleration of the earth. This acceleration is the trigger for a Noahic Event. Responsibility for recurrence of Noahic Events rests with these trigger events at the surface, not with the core. The full nature of such trigger events is far from understood at present, though the Balleny Islands example implies a fundamentally volcanic process. What we know about volcanic eruptions is that there have been very many throughout earth history. There appears to be no obvious reason why the Balleny Islands eruption should have been a singular event.

The trigger event causes the inner core to be nudged off center, ultimately to collide with the mantle due to gravitational collapse. Core-wide fluid currents generated in the process keep the inner core pinned to the mantle for a time. But, eventually, collapse ceases and a rebound occurs as overly compressed matter begins to decompress. Core-wide circulation slows, allowing the inner core to "sink" gradually back to center, where it is able to support the earth through whatever residual oscillations may follow. Thus the mechanism appears to reset itself automatically.

This cannot go on forever, of course. Gravitational energy is lost to friction in several ways each time a Noahic Event is triggered. Consequently, the earth will have shrunk a little after each Noahic Event. But the system is hardly out of gravitational energy yet.

While many details regarding the mechanism of Noah's Flood remain to be filled in, it presently appears that a Noahic Event could happen at any time, including immediately following a prior Noahic Event.

Argument 2

Geochronology finds that the age of Iceland and the age of the Balleny Islands both greatly exceed the interval of five and a half thousand years which has transpired since the Flood. This says that these volcanic hotspots came into being long before the Flood. Yet their antipodal orientation and their hotspot character both find natural explanation as results of a Noahic Event, as we have previously seen. This implies that the Balleny Islands triggered at least one other Noahic Event prior to Noah's Flood.

Evidently, then, a single trigger site can erupt multiple times and trigger multiple Noahic Events—hardly surprising for a fundamentally volcanic process. And this implies that Noah's Flood was not a singular event.

Argument 3

High-energy, antipodal surface features now appear to be signatures of Noahic Events. Instances of high-energy, antipodal surface features besides the Balleny Islands – Iceland pair have been identified. Jonathan T. Hagstrum has compiled a list of such antipodal hotspot pairs. He reports:[1]

> Of 45 'primary' hotspots found in most hotspot compilations 22 (49%) form antipodal pairs within observed hotspot drift limits ($\leq$ 20 mm/yr). ...All hotspot pairs include at least one oceanic hotspot, and these are consistently opposite those hotspots related to large igneous provinces (LIPs) and continental volcanism.

The Balleny Islands – Iceland antipodal pair is included in Hagstrum's list. While some of the remaining 10 antipodal hotspot pairs may be coincidental, they cannot all be. In each instance, the oceanic hotspot would be the trigger site. It is the analog of the Balleny Islands. The antipodal LIP hotspot is a site where there has been a very large volume of lava poured out onto the surface. It is the analog of Iceland. The trigger in each instance evidently leads to flood basalt volcanism on the opposite side of the globe due to energetic collision of the inner core with the underside of the mantle.

If true antipodal hotspot pairs are smoking-gun evidences of Noahic Events, then Noah's Flood was not a singular event.

Argument 4

To trigger a Noahic Event, it is necessary, according to the mechanism of the Flood presented in this volume, to accelerate earth in space. Any acceleration of earth in space will change earth's orbit about the sun (Figure 40.1). Changes in earth's orbit about the sun have obvious potential to alter earth's climate.

Additionally, the lopsided distribution of mass during a Noahic Event—with the inner core off to one side of the core, and the oceans and atmo-

[1] Jonathan T. Hagstrum, "Antipodal hotspots and bipolar catastrophes: Were oceanic large-body impacts the cause?," *Earth and Planetary Science Letters* 236 (2005): 13–27.

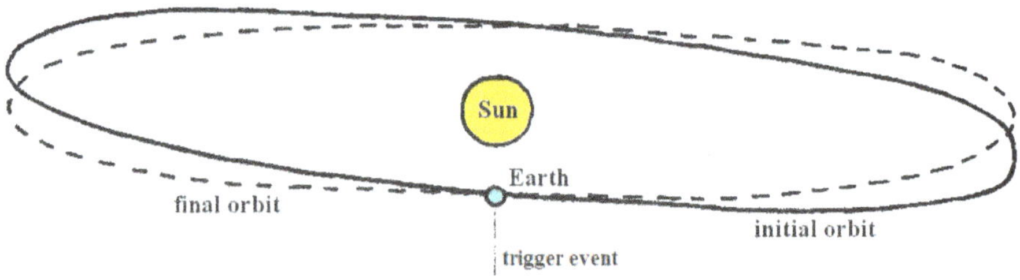

Figure 40.1: Sketch (not to scale) illustrating one way a trigger event at the South Pole can alter earth's (counter-clockwise) orbit about the sun. In this particular case, earth's tilt with respect to its orbital plane will be altered (since the orientation of its spin axis in space will remain the same). In the general case, the effect on earth's orbit will depend on factors such as the location of the volcanic trigger event, the time of year, and the duration and intensity of the acceleration. Notice, for example, that in the case shown, earth's tilt would change in the opposite sense for an equivalent trigger event a half-year later.

sphere following suit at the surface—will cause the rotation axis of the earth to precess. This also has potential to alter climate.

The Noahic Event which produced Noah's Flood five and a half thousand years ago, in the mid-Holocene, does seem to have changed earth's climate:[2]

> The Standard$_{5\times5}$ reconstruction exhibits ~0.6°C of warming from the early Holocene (11,300 yr B.P.) to a temperature plateau extending from 9500 to 5500 yr B.P.. This warm interval is followed by a long-term 0.7°C cooling from 5500 to ~100 yr B.P.

That is, Holocene climate was warmer pre-Flood (particularly in the Northern Hemisphere) than it was post-Flood. (The first half of the Holocene is often called the Holocene Climatic Optimum.) Cooling began 5500 years ago, coincident with the Flood.

Another clear example of this climate change is the desiccation of the Sahara post-Flood. Prior to the Flood, much of the Sahara was a savanna inhabited by humans and animals. Today it is a desert wasteland.

[2]Shaun A. Marcott et al., "A Reconstruction of Regional and Global Temperature for the Past 11,300 Years," *Science* 339 (8 March 2013): 1198.

Lacking any other obvious causative factor, scientists frequently fall back on Milankovitch Cycles (millennia-scale changes in earth's orbit about the sun) to explain such climate changes, and this is true of the observed desiccation of the Sahara. But Milankovitch Cycles are calculated on the assumption that earth's orbit is not disturbed in any way—that it is always subject only to the same gravitational forces acting on it today. Noahic Events obviously violate this assumption (Figure 40.1).

How severely Noahic Events violate this assumption will become clear only when more is known about them. But Milankovitch Cycles and Noahic Events make very different predictions regarding the rapidity with which the Sahara might be expected to transition from a moist to a dry climate. This makes it possible to begin to foresee something of the relative importance of these two explanations for the mid-Holocene change in climate even at this early stage of relative ignorance. Milankovitch Cycles result in millenia-scale changes in earth's orbit while Noahic Events, as has been shown, are expected to result in year-scale changes in earth's orbit. By this rapidity-of-action test, Noahic Events are clearly the favored explanation. Marine sediment cores show that the transition from savanna to desert took place rapidly (decades to centuries) over much of northern Africa.[3]

Thus Noah's Flood seems to fulfill theoretical expectations of a cause-and-effect relationship between Noahic Events and rapid climate change.

Once it is understood that Noahic Events have potential to alter earth's climate, then otherwise unexplained climate changes become another possible indicator of Noahic Events. And there are, at present, many unexplained changes in earth's climate.

The most conspicuous unexplained climate changes are those associated with the ice ages. For example, at present we are in an inter-glacial period. The transition from glacial period (Younger Dryas) to the present inter-glacial (Holocene) happened about 11,500 years ago. Like the transition into the present-day dry Sahara, the transition to present-day inter-glacial happened suddenly, on the order of a decade.

It seems likely at this stage that unrecognized Noahic Events may be the cause of some previously unexplained climate changes. If so, then Noah's Flood was not a singular event.

---

[3]See for example: Jessica E. Tierney and Peter B. deMenocal, "Abrupt Shifts in Horn of Africa Hydroclimate Since the Last Glacial Maximum," *Science* 342 (15 November 2013): 843–846.

Argument 5

A longstanding question in geophysics is what causes plate tectonics.[4] The slow movement of plates over the face of the globe is a fact of modern earth science. But the root physical cause of this motion—the driving force behind plate tectonics—continues to be debated.

Preference seems to rest at present with the idea that the weight of a subducted plate edge drags a plate across the surface. But it is recognized that even if this is true, some other mechanism is needed to get plate motion started to form subduction zones in the first place.

Why should the earth have plates rather than just a continuous solid shell? Why should these plates be actively moving about? The idea that Noahic Events have been happening repeatedly throughout earth's history provides simple, intuitive answers to these otherwise difficult questions.

To see how this works, imagine a hollow ceramic ball the size of Earth which has a stainless steel ball the size of the inner core untethered inside it. (By analogy with the earth, the stainless steel ball is in a viscous fluid which fills the hollow ceramic ball, but this is not very important in the present context.) In a Noahic Event, the stainless steel ball collides energetically with the interior wall of the ceramic ball. What will this do? Fracturing of the ceramic ball is the most obvious expectation.

For a strong enough collision, or for multiple weaker collisions, it is possible to fracture the ceramic ball into numerous pieces. These pieces will not fall apart from one another because gravity holds them together. Call these pieces "plates" and you have a basic answer to why it is that the earth should have plates to begin with.

Next, add in the inevitable squeezing together of these plates due to gravitational collapse of the earth during the compression phase of a Noahic Event. As earth's radius shrinks, its circumference must also shrink, and this provides both the energy and the opportunity needed for some plates to begin to be driven beneath other plates. Thus Noahic Events provide an initiation mechanism for plate subduction.

Finally, add in rebound of the earth. During the decompression phase of a Noahic Event, the globe expands back to roughly its previous size.

[4]Many good primers on plate tectonics can be found on the Internet. See for example: http://www.leitzelcenter.unh.edu/geo-teach/pdf/ESST2008/Johnson-MartinetalPlateTectonics.pdf.

Plates now have a reason to pull apart from one another. But if one edge of a plate has been driven beneath a neighboring plate, it is likely to be somewhat stuck there. The plate's opposite edge is then likely to pull away from its neighboring plate. Thus Noahic Events provide a simple explanation of divergent plate boundaries (i.e., mid-ocean ridges).

Said simply, plate tectonics is the expected outcome in the broken, upper, solid shell of the earth of repeated cycles of contraction and re-expansion of the whole earth. Thus plate tectonics appears to testify to repeated Noahic Events throughout earth's past. And this is just another way of saying that Noah's Flood was not a singular event.

Argument 6

The cause of earth's magnetic field has long been researched, but is still not understood.[5]

Earth's magnetic field is believed to be due to electric currents in earth's core. It is an electromagnetic field, not a permanent magnetic field.

The electrical currents responsible for the magnetic field are presently believed to be sourced by fluid motions in the outer core. Outer core fluid is believed to be principally molten iron, which is electrically conductive. Interaction between this moving, conducting fluid and existing magnetic field lines in the core is believed to generate currents which produce the magnetic field. Thus the earth's magnetic field is believed to be sourced via a natural electric generator operating in earth's core. This natural generator is called the "geodynamo."

According to present scientific consensus, the geodynamo is due to outer core fluid motions driven by convection. Convection is believed to be fueled in part by gravitational potential energy which is released by precipitation of iron from outer core fluid onto the solid inner core as the core slowly cools.

Substantial revision of this present consensus view seems required by what has been learned about the core from Noah's observations of the Flood in the present book. Notice, to begin with, that the idea of outer core fluid currents driven by convection is called into question.

---

[5]Many fine primers on geomagnetism can be found on the Internet. See for example: http://www.geomag.bgs.ac.uk/education/reversals.html, http://www.scientificamerican.com/article/what-causes-the-periodic/, and http://www.scientificamerican.com/article/why-does-the-earths-magne/.

The idea of convection is rooted in an assumption of uniform, undisturbed cooling of the core over millions of years. Noahic Events falsify this assumption. They say that the core gets forcibly stirred from time to time. Thermal gradients and density stratification of outer core fluid would both be lost in the case of such forced mixing, and core temperatures would necessarily rise for a time as a result of friction within the stirred outer core fluid. The assumption that the core cools monotonically over millions of years is evidently incorrect.

The idea that convection in the core drives the geodynamo is further called into question by Noah's observations implying that the viscosity of the outer core is very large, as previously discussed (page 243). Outer core viscosity is typically assumed to be on the order of 10^{-2} Pa-s in discussions of the geodynamo.[6] But we have seen that Noah's observations imply a viscosity during the Flood which probably exceeded 10^{16} Pa-s. It appears that the outer core may be too viscous to allow the magnitude of convection needed to fuel the geodynamo.

Regardless, Noahic Events point in another direction entirely. They imply that earth's magnetic field is due to a totally different sort of electromagnetic machine.

Most of the concepts and processes which have come to be associated with the geodynamo are not applicable to this new machine. While this new machine may also be called a geodynamo, in the sense that any device which converts energy of motion to electrical energy is a dynamo, I will call it a "geogenerator" in an effort to avoid confusion between the old and the new machines.

The geogenerator is fueled by contraction of the core during a Noahic Event, not by convection of fluid within the core. It is a simple Faraday induction machine.

Faraday's law of induction states that the electric current induced in a loop of wire by a magnetic field is proportional to the rate of change of total magnetic flux threading the loop.

Think of the core as made up of a large number of conducting loops or rings. These rings are all aligned perpendicular to and concentric with the north–south geomagnetic axis. They are all threaded by magnetic field lines from earth's magnetic dipole field. When the core contracts during a Noahic Event, these rings all become smaller. Focus on one ring,

---

[6]See for example: Paul Melchior, *The Physics of the Earth's Core*, (Oxford: Pergamon Press, 1986).

say the equatorial ring. Magnetic field lines which had been inside this ring but out near its radius before contraction will, after contraction, be outside the ring.

Thus it is seen that the total magnetic flux inside each of these conducting loops will change with time during the Flood. This change of flux with time induces new currents in all the rings in accordance with Faraday's law of induction. These new currents generate new magnetic field.

As a simple check of this new mode of generating earth's magnetic field, I first approximated the source of the pre-contraction magnetic field as being due to a single conducting ring. I then asked whether the magnitude of the current induced in this ring by contraction of the core would exceed the magnitude of the normal (i.e., pre-contraction) current in the ring, responsible for maintaining the earth's normal magnetic field.

This time I was able to carry out the calculation using pen and paper. I took the radius of the ring to be half the normal core radius. I took the electrical conductance of the ring to be the conductivity of the core times half of the cross-sectional area of the core divided by the circumference of the ring. For a constant, non-zero magnetic field in the core, I found that the ratio of the induced current, i, to the normal current, i_0, is given by the equation:

$$\frac{i}{i_0} = \frac{\mu\sigma\delta V}{16R\delta t} \tag{40.1}$$

where μ is the permeability of the core (taken to be 1×10^{-6} henries/meter), σ is the electrical conductivity of the core (taken to be 5×10^5 siemens/meter), R is the radius of the core (3.5×10^6 meters), and δV is the total change in volume of the core during the time interval δt. The time interval of interest in the present case is the 40 days (3.5×10^6 seconds) of the Flood during which the inner core was moving to the mantle. The actual contraction of the core during this time interval is not known at present, but a minimum δV of 5.7×10^{16} cubic meters has previously been calculated (page 254). Use of these values gives:

$$\frac{i}{i_0} > 1 \times 10^2. \tag{40.2}$$

Thus, according to this calculation, contraction of the core induces a current at least 100 times larger than the normal current. This shows immediately that the geogenerator is able to produce currents large enough to explain earth's magnetic field.

In fact, the induced current may seem at first to be too large, giving a resultant magnetic field which is much too large, but this is not the case. This is the steady state current. Since the time constant for changes to earth's magnetic field is 10^4 years or more, only a small percentage of the steady state current will be achieved during the relatively short-lived contraction phase of a Noahic Event. In addition, a similarly large (steady state) current will be induced in the opposite direction during the rebound stage of a Noahic Event. Thus the size of the residual current (and residual magnetic field) once the Noahic Event has run to completion will be the difference between these two competing induction processes. It will obviously be much smaller in magnitude than the steady state current resulting from either individual process.

Earth's magnetic field is known to be very much older than the Flood. Thus, if the earth's magnetic field is due to a geogenerator activated by core contraction during Noahic Events, then Noah's Flood was not a singular event.

Conclusion

Noah's Flood appears to have been just the most recent instance of a previously unrecognized class of devastating earth catastrophes.

Chapter Forty-One

The Recent Frequency of Noahic Events

Once it has been understood that Noah's Flood was just the most recent expression of the Noahic Event category of earth catastrophes, the question of present-day risk of recurrence presents itself. To answer this question, knowledge of the average frequency of Noahic Events in recent times is needed.

Despite considerable investment of time and energy into this problem, I still have no certain solution to it. There are, at this stage, too many unknowns, and these will require the investment of many hundreds of hours by earth's scientists to resolve. But two methods of estimating this frequency yield disquieting results.

Geomagnetic Field Reversals

Measurements of remnant magnetism in rocks have shown that earth's magnetic field reverses polarity from time to time. If earth's magnetic field is attributed to a geodynamo, then this observation demands that the geodynamo be capable of spontaneously reversing its field polarity. Spontaneous polarity reversal is not a necessary property of the geogenerator. Instead, it seems likely that polarity reversals are intrinsic to the nature of field generation during a Noahic Event.

It has already been seen that two competing induction processes are active during a Noahic Event: the first during the core compression phase, inducing a large steady state current in one direction, and the second during the decompression phase, inducing a large steady state current in the opposite direction. Thus the residual field results from subtraction of two, similar large numbers. Whether the result will be positive or negative in each instance will likely have a random character. When more subtle effects are added in to the overall magnetic field production process, such as the forced mixing of the electrically conductive core during a Noahic Event, the idea of the polarity of the resultant field being random seems strengthened.

Thus earth's magnetic field appears to be reborn with each Noahic Event. How much "memory" of its previous existence is retained in each rebirth, influencing its new polarity, is not clear at this stage. But the important point at present is only that reversals are expected to be due to rebirth of the field in a Noahic Event. This makes geomagnetic field reversals to be another signature of Noahic Events, and this allows the chronology of field reversals worked out by the geomagnetists to be used to estimate the minimum frequency of Noahic Events.

This chronology clearly reveals that reversals happen at random. About a dozen reversals have been measured in samples having ages spanning the past 3 million years, yielding an average reversal frequency at present of about one per 250,000 years.

This is less comforting than it might seem at first because it represents only a *minimum* frequency estimate for Noahic Events. Not every Noahic Event is expected to reverse the polarity of the field. If memory effects are significant, then polarity reversals may have a low probability.

Most disturbing, however, is the fact that reversals are known to have occurred very closely together. The Laschamp event, for example, shows two reversals in less than a thousand years.[1]

> The period of reversed magnetic field was $\sim$ 440 years, with the transition from the normal field lasting $\sim$ 250 years.

If geomagnetic field reversals are signatures of Noahic Events, then they support the conclusion of the previous chapter (page 290) that Noahic Events can happen at any time, including immediately following a previous Noahic Event.

[1] http://en.wikipedia.org/wiki/Laschamp_event

Earth's Climate

Another method of estimating the recent frequency of Noahic Events is based on the observation discussed in the previous chapter that Noahic Events have potential to alter earth's climate. Unfortunately, unexplained changes in earth's climate appear to be relatively frequent.

If the cause of the transition from Younger Dryas to Holocene which happened about 11,500 years ago, mentioned previously, was perturbation of earth's orbit due to a Noahic Event, then a frequency for recent times of roughly one Noahic Event per 6000 years results.

But this is hardly the end of relatively recent unexplained climate transitions. There is, for example, also the transition into the Younger Dryas, which happened suddenly about 13,000 years ago. If this transition was also caused by a Noahic Event, then the frequency would be three events in 13,000 years, for an average of roughly one per 4,000 years.

But then there is also the so-called 8.2 kiloyear climate event. Here again there were rapid transitions into and out of a markedly colder climate in the Northern Hemisphere. If these two transitions are included, the relatively recent frequency becomes roughly one event per 3,000 years.

And these are hardly the end of the possibilities. They are just the most conspicuous examples. Given the relative subtlety of the effect on climate of the most recent Noahic Event (i.e., Noah's Flood), there is clearly warrant to include less pronounced climate perturbations in the list of possibilities. But I think the point has been adequately made. This method appears to yield an average period of a few thousand years, making the interval of 5500 years which has transpired since the Flood seem already overlong.

Conclusion

The long history of civilization, stretching back thousands of years to the building of the pyramids and beyond, has seemed an assurance that our planet can be counted on as an essentially stable base for the development of civilization into the foreseeable future. This is now called into question. Noahic Events may be distressingly frequent. It is possible that the rise of modern, technologically advanced civilization on earth may have been made possible by an unusual absence of Noahic Events since Noah's Flood. The present safety of civilization is necessarily called into question.

Chapter Forty-Two

Conclusion to Part 6

A correct understanding of the true nature of Noah's Flood instigates a thorough rethinking of established wisdom in a variety of fields. Topping the list is the need to rethink the safety of earth's inhabitants.

The understanding of Noah's Flood presented in this volume provides an entirely new window into the workings of earth's core. Peering through this new window at this still early stage, we are able to see things about the behavior of the core which have never before been seen. We see, among other things, that the inner core's residence on center is not stable. It appears to be more like a ball between strongly compressed horizontal springs, ready to go flying if slightly displaced vertically. In this new view, Earth appears as a global catastrophe perpetually waiting to happen.

The inner core was nudged from center five and a half thousand years ago by volcanic propulsion of the earth at Balleny Islands, and the catastrophe of Noah's Flood chronicled in Genesis was the result. Evidence from a variety of sources argues that there have been other suitable pushes triggering other Noahic Events in the past. What the probability may be of a suitable push in the present or near future is not yet known, but an average frequency of such events at present of every few thousand years seems a possibility.

Clearly, the blindness to the historical reality of Noah's Flood which has characterized contemporary science can no longer be indulged in or tolerated by men and women of science having insight and integrity. The facts are now clear. The ancients have told us that civilization was all but snuffed out by a catastrophe of stupendous proportions. We now understand what they were talking about. They were talking about a previously unsuspected type of catastrophic behavior of our planet which makes even the most dire global warming scenario look like a Sunday-school picnic and which now appears intrinsic to earth with its two-phase core. This geophysical process can no longer be ignored.

Epilogue

CHAPTER FORTY-THREE

RADIOCARBON DATING NOAH'S FLOOD

Some years ago, I attempted—unsuccessfully—to radiocarbon date Noah's Flood.[1] My purpose was to check the modern 3520±21 B.C. biblical chronology date for the Flood. Having discovered that traditional biblical chronology had left out a full millennium prior to the first millennium B.C., reassurance that the newly corrected chronology was now sound was obviously desirable.

To this end, tiny samples of the ancient Cave of the Treasure, Israel, reed mat were obtained and submitted for AMS radiocarbon dating. I thought the mat might have been fabricated just prior to the Flood. If so, dating the mat would date the Flood.

Multiple AMS radiocarbon dates on the mat turned out to be scattered and overall too old for the Flood. So my attempt at a direct radiocarbon date for the Flood ended in failure.

The recent two-volume book *Gathering Time: Dating the Early Neolithic Enclosures of Southern Britain and Ireland*[2] quite inadvertently does what my deliberate attempt failed to do. It inadvertently radiocarbon dates the Flood.

---

[1]Gerald E. Aardsma, "New Radiocarbon Dates for the Reed Mat from the Cave of the Treasure, Israel," *Near East Chronology: Archaeology and Environment. Proceedings of the 17th International* ^{14}C *Conference*, ed. Hendrik J. Bruins, I. Carmi, and E. Boaretto *Radiocarbon* 43.3 (2001): 1247–1254.

[2]Alasdair Whittle, Frances Healy and Alex Bayliss, *Gathering Time: Dating the Early Neolithic Enclosures of Southern Britain and Ireland* (Oxford: Oxbow Books, 2011).

To avoid sullying any reputations or damaging any academic careers, let me hasten to clarify that *Gathering Time* has nothing to do with the Flood as far as its authors and contributors are concerned. *Gathering Time* is a truly fine work, in the best tradition of thorough, objective, well-funded scientific investigation. It is wholly oblivious of the close connection of its subject area to Genesis—which is partly why it does such a fine job of checking the modern biblical chronology date of the Flood. Oblivion guarantees objectivity.

Gathering Time concerns itself with causewayed enclosures of southern Britain and Ireland. Enclosures are monumental earthworks of ancient origin. Think of a central, more or less flat area a few acres in size. Now surround it with a deep ditch. Heap the dirt from the ditch up as a hill paralleling the ditch to yield a rampart. Leave the ditch undug in a few places so occupants can enter and leave the enclosed area on level ground and you have a basic causewayed enclosure.

More accurately stated, *Gathering Time* concerns itself with *dating* causewayed enclosures of southern Britain and Ireland. For this purpose, it uses many hundreds of radiocarbon dates on over 35 enclosures. It then uses Bayesian statistical analyses to answer such questions as when these enclosures started to be built and when they stopped being built.

They started being built at various times spanning several centuries prior to the Flood. And, as you may by now have anticipated, they stopped being built coincident with Noah's Flood:[3]

> Intensive circuit building [i.e., ditch digging] continued into the last quarter of the 36th century cal BC.

The "last quarter of the 36th century cal BC" begins 3525 B.C. and ends 3500 B.C. The biblical chronology (3σ) date range for the Flood is 3541 B.C. to 3499 B.C. The agreement between these two date ranges, independently calculated for the same event, is a bit breath-taking for such a remote time.

The agreement between these two independent estimates of the date of the Flood is illustrated graphically in Figure 43.1. The red and blue lines show "proportion of total ditch length excavated in each 25-year

[3]Alex Bayliss, Frances Healy, Alasdair Whittle and Gabriel Cooney, "Neolithic Narratives: British and Irish enclosures in their timescapes," *Gathering Time: Dating the Early Neolithic Enclosures of Southern Britain and Ireland*, Alasdair Whittle, Frances Healy and Alex Bayliss (Oxford: Oxbow Books, 2011), 698.

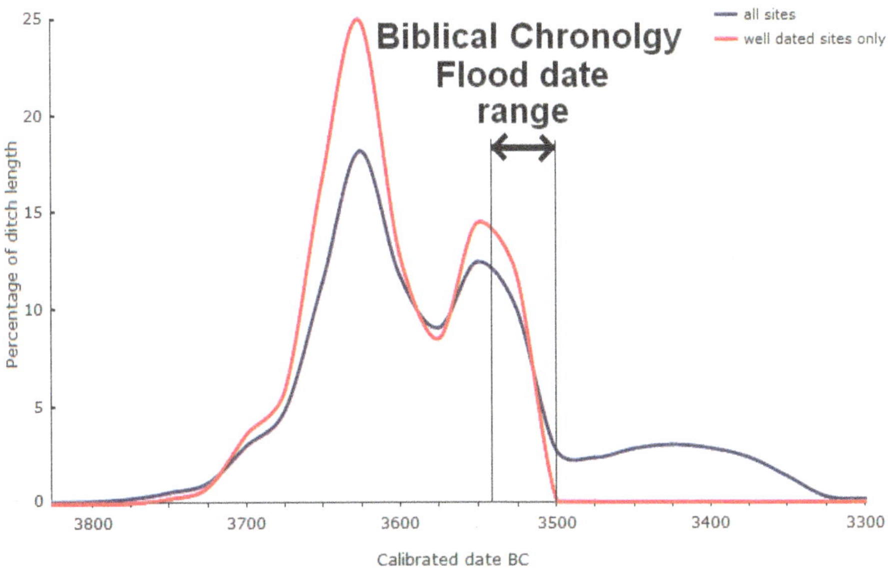

Figure 43.1: Percentage of ditch length versus calendar date B.C. for 38 ditches (many enclosures have more than one surrounding ditch). Curves are drawn in the original *Gathering Time* figure from 25-year interval histograms. I have added the vertical black lines showing the 3σ time window for the date of Noah's Flood given by modern biblical chronology. [Figure adapted from Alex Bayliss, Frances Healy, Alasdair Whittle and Gabriel Cooney, "Neolithic Narratives: British and Irish enclosures in their timescapes," *Gathering Time: Dating the Early Neolithic Enclosures of Southern Britain and Ireland*, Alasdair Whittle, Frances Healy and Alex Bayliss (Oxford: Oxbow Books, 2011), Figure 14.20, 696.]

period" according to the original figure's caption. The red line is for "well dated sites only" and the blue line is for "all sites." Both curves show a dramatic reduction in enclosure construction coincident with the new biblical chronology date for the Flood.

Improving the Check

Both the red and the blue curves are drawn assuming continuity. This assumption is now known to be false. The Flood came, introducing a profound discontinuity.

Said more quantitatively, the highest point in the United Kingdom (according to Wikipedia) is 1344 meters above mean sea level. The angular distance from Iceland to the farthest shore of the United Kingdom is less than 18°. From Figure 34.2 (page 261) [or, more easily and accurately, from the table in Appendix E (page 339)], the Flood is found to have attained a depth of 7850 meters (i.e., nearly 8 kilometers) at 18° from Iceland. Thus the highest point in the United Kingdom was at least 6,500 meters (i.e., over 6 kilometers, or over 4 miles) under water during the Flood. From Figure 16.3 (page 119), it is found that this was the case for at least 110 days. Under these circumstances, it is clear that excavation would suddenly have ceased altogether. It may further reasonably be inferred that sufficient population base would not have existed for some centuries following the Flood for excavation of any of these ditches to have been resumed (even assuming there remained any reason to want to go to the enormous effort required to excavate them). This knowledge may be used to improve this radiocarbon check of the date of the Flood.

To begin with, for the present purpose, the blue curve may safely be discarded altogether. It shows resumption of excavation immediately following the Flood, which is clearly in error.

The red curve, with its more stringent selection of radiocarbon dates, does not suffer this problem. But it needs to be corrected for its mistaken assumption of continuity. The red curve should not drop smoothly to zero. It should drop suddenly to zero when the Flood happens.

To correct this problem, it is necessary to go back to the histogram data from which the red curve was drawn. I have graphically reconstituted the underlying histogram data to the best of my ability in Figure 43.2. Four 25-year bins have been reconstituted, covering the 36th century B.C. Bin "a" represents a lull in excavation, the cause of which

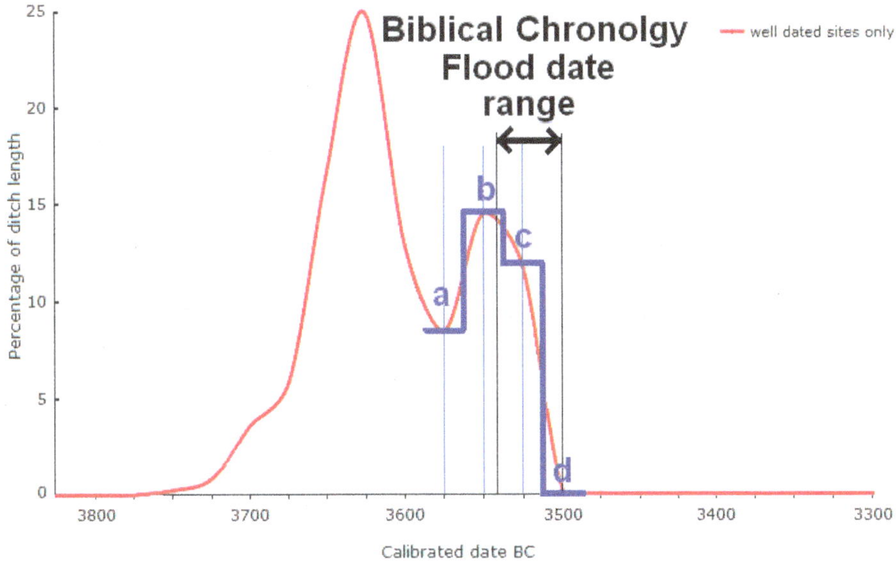

Figure 43.2: My graphically reconstituted 25-year data for the 36th century B.C. [Figure adapted from Alex Bayliss, Frances Healy, Alasdair Whittle and Gabriel Cooney, "Neolithic Narratives: British and Irish enclosures in their timescapes," *Gathering Time: Dating the Early Neolithic Enclosures of Southern Britain and Ireland*, Alasdair Whittle, Frances Healy and Alex Bayliss (Oxford: Oxbow Books, 2011), Figure 14.20, 696.]

is unknown. Disease may have swept the country, reducing human populations, or widespread famine may have been the cause, or human aggression, or something else entirely. Bin "b" shows strong recovery and resumption of excavation. Bin "c" shows an apparent decline in excavation, and bin "d" shows no excavation at all.

It is immediately clear that the Flood must have come in bin "c." That is why excavation had ceased in bin "d."

Thus the radiocarbon date for the Flood using these causewayed enclosures data is 3525±12.5 B.C. (3σ), for this is the interval spanned by bin "c."

Thus the biblical chronology date for the Flood and the radiocarbon date for the Flood from these causewayed enclosures are in close agreement. The modern biblical chronology date for the Flood appears to be sound.

While this result formally concludes this check, it is appropriate to go a bit further and finish the reworking of Figure 43.1 (page 309) resulting from recognition of the historical reality of Noah's Flood. Figure 43.3 shows the final result of the reconstituted histogram data with bin "c" corrected for its narrower width due to the advent of the Flood 3520 B.C. The original red curve (Figure 43.1, page 309) gave the impression that construction of enclosures slumped a second time beginning 3550 B.C., and ground slowly to a halt over the next 50 years. This impression resulted entirely from the mistaken assumption of continuity. Figure 43.3 shows that in actual fact there was no second slump. Construction continued to strengthen following the first slump, right up until the Flood came and catastrophically terminated enclosure construction altogether.

Best Estimate of the Date of Noah's Flood

Averaging the biblical chronology date of the Flood with this independent radiocarbon date yields 3522±12 B.C. (3σ) as the best current estimate of the date of the Flood. The ±12 years uncertainty implies that this result should be rounded to the nearest decade when quoted without accompanying error estimate, as I have done, for example, in the title of this book.

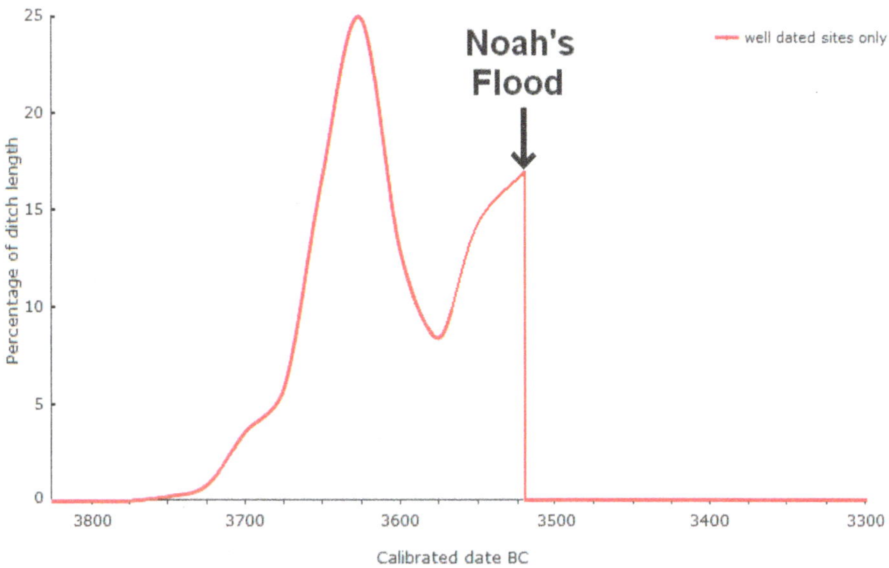

Figure 43.3: My corrected ditch digging intensity versus calendar years resulting from knowledge that the Flood happened 3520 B.C. [Figure adapted from Alex Bayliss, Frances Healy, Alasdair Whittle and Gabriel Cooney, "Neolithic Narratives: British and Irish enclosures in their timescapes," *Gathering Time: Dating the Early Neolithic Enclosures of Southern Britain and Ireland*, Alasdair Whittle, Frances Healy and Alex Bayliss (Oxford: Oxbow Books, 2011), Figure 14.20, 696.]

Appendices

Appendix A

Critical Height for a Submerged Ice Sheet to Remain Frozen to Its Bed

We wish to know the critical thickness of an ice sheet below which it is safe to assume the sheet will remain attached to its bed when submerged by water.

Figure A.1 shows the basic problem. If the force of buoyancy, F_b, acting on an ice column due to the presence of water is greater than the weight of the ice column, F_g, plus the force due to the ultimate tensile strength of the ice, F_t, holding it to its bed, then the ice will break away from the bed and float. Otherwise it will stay submerged.

There will be no buoyant force lifting an ice column from its bed so long as water can be kept from getting under the ice, in which case the ice sheet will stay submerged indefinitely.

In real life, however, it seems impossible to keep water from getting under (or into) the ice. Real ice sheets generally contain crevasses and also ice caves extending to the bed. In addition, even in the (idealized) case of a completely solid sheet of ice, it seems inevitable that water molecules, under the pressure of the overlying water column, will penetrate the ice fabric and cause it to fracture whenever the pressure they exert internal to the ice exceeds the tensile strength of the ice. It thus appears that deep enough water will inevitably exert a buoyant force on submerged ice sheets.

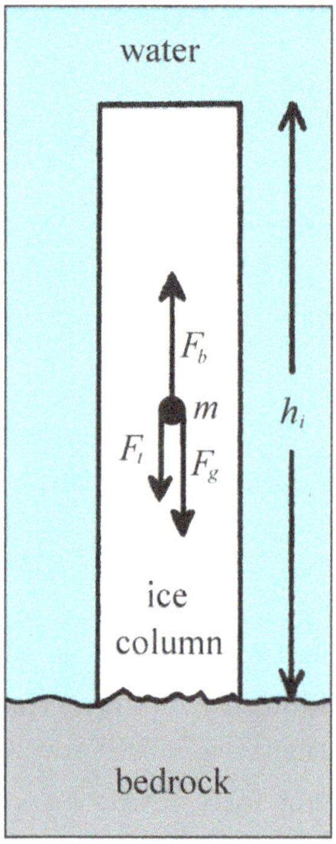

Figure A.1: Forces acting on a submerged ice column frozen to its bed.

Assuming that buoyancy is active, lifting the ice, the force of buoyancy on the ice is, according to Archimedes' principle, just equal to the weight of the displaced water. Let ρ_w represent the density of sea water, V_i represent the volume of the column of ice shown, and g represent the acceleration due to gravity. Then

$$F_b = \rho_w V_i g. \tag{A.1}$$

Let ρ_i represent the density of the ice. Then

$$F_g = \rho_i V_i g. \tag{A.2}$$

At the breaking point, the buoyant force will be equal to the sum of the other two forces.

$$F_b = F_g + F_t \tag{A.3}$$

Substituting for F_b and F_g gives

$$\rho_w V_i g = \rho_i V_i g + F_t. \tag{A.4}$$

Rearranging gives

$$V_i g(\rho_w - \rho_i) = F_t. \tag{A.5}$$

Let the cross-sectional area of the column of ice be A_i and its height be h_i. Then

$$V_i = A_i h_i \tag{A.6}$$

Solving these last two equations for h_i yields the desired result.

$$h_i = \frac{F_t}{A_i g(\rho_w - \rho_i)}. \tag{A.7}$$

Substituting values for the quantities on the right side of this equation yields the critical height of a column of ice. An ice sheet which is on average thicker than this critical height will be expected to break away from its bed and float, while thinner ice sheets will remain frozen to their beds and be submerged.

The quantity $F_t/(A_i g)$ (the ultimate tensile strength of ice) has been measured by a number of investigators. W. F. Weeks found a value of $29.6 \pm 8.5\ \mathrm{kg/cm^2}$ for fresh-water ice produced in the laboratory.[1] Butkovich

[1] W. F. Weeks, "Tensile Strength of NaCl Ice," *Journal of Glaciology* 4.31 (1962): 46.

found values within this range for measurements on glacier ice[2] as did Langleben and Pounder for sea ice[3].

The density of ocean water varies slightly with temperature, salinity, and pressure. For the present case, a value for ρ_w of about 1033 kg/m^3 seems most appropriate.[4]

Glacier (or ice sheet) ice contains air bubbles, making it less dense than air-free ice. Icebergs are pieces of ice sheets which have broken off into the ocean. They should provide a representative value of ice density for ice sheet ice. Pickard and Emery give the density of iceberg ice as "about 900 kg/m^3."[5] Given the large uncertainty in the ultimate tensile strength of ice cited above, this estimate should be adequate for the present calculation.

Substitution of these values into Equation A.7 yields a critical height of about 2200±600 meters (7,300±2,100 feet). For heights less than this a submerged ice sheet frozen to its bed is guaranteed to remain submerged.

---

[2] W. F. Weeks, "Tensile Strength of NaCl Ice," *Journal of Glaciology* 4.31 (1962): 34 (Table I).

[3] M. P. Langleben and E. R. Pounder, "Arctic Sea Ice of Various Ages: I. Ultimate Strength," *Journal of Glaciology* 5 (1964): 95 (Table I).

[4] George L. Pickard and William J. Emery, *Descriptive Physical Oceanography: an Introduction*, 4th (SI) enlarged edition (New York: Pergamon Press, 1982), 17–21.

[5] George L. Pickard and William J. Emery, *Descriptive Physical Oceanography: an Introduction*, 4th (SI) enlarged edition (New York: Pergamon Press, 1982), 184.

Appendix B

Dating Devon Island Ice Core D72

The chronological data needed to date ice from the Devon Island ice sheet are the measured annual layer thicknesses at various depths along a given borehole. These were determined by the original researchers using the ice cores which were taken from the ice sheet. Seasonal variations in measured microparticle (i.e., dust) concentrations were used as the means of distinguishing one year from the next—annual snow layers soon lose all visual distinctiveness because of compression by overlying layers and conversion to solid ice.

A continuous series of microparticle measurements from the top of the ice core to its base would be very nice at this point. Such a series would make it possible simply to count annual layers of ice from the top down to 13 meters above bedrock. This would provide the simplest method of ascertaining the date of the ice at 13 meters. However, while this method provides for simple chronology, it requires a very large effort in the laboratory. As a result, no such continuous series exists.

To discern yearly oscillations in microparticle concentrations requires separate measurements on at least two samples from each annual layer. This means that to date back to five thousand years ago by counting annual microparticle oscillations, the microparticle concentrations of at least ten thousand samples of ice would need to be measured in the laboratory. (In actual practice, of course, four or more samples per year would need to be measured to resolve the annual signal adequately.)

Rather than go to the enormous amount of work and expense re-

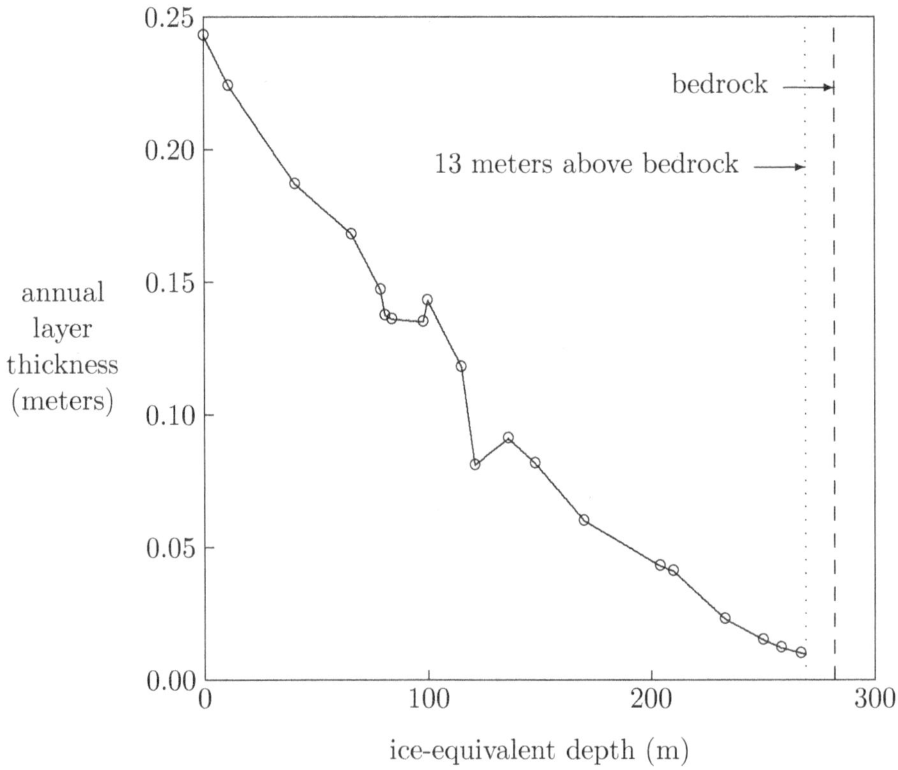

Figure B.1: Annual layer thickness versus ice-equivalent depth for Devon Island ice core D72. The circles represent actual measurements.

quired to measure microparticle concentrations throughout an entire ice core, it is easier and quicker (though less accurate, of course) to measure microparticle concentrations in detail at a relatively small number of depths along the core and interpolate these measured values. This is what has been done with the Devon Island cores.

Figure B.1 shows a graph of measured annual layer thicknesses versus depth for the Devon Island ice core D72.[1] I have interpolated the measured data points with straight lines, and linearly extrapolated the last two points to the 13 meter-above-bedrock depth.

The total length of ice and snow in the D72 core was 298.9 meters,

[1]N. Reeh and W. S. B. Paterson, "Application of a Flow Model to the Ice-divide Region of Devon Island Ice Cap, Canada," *Journal of Glaciology* 34.116 (1988): 62, Table I.

as mentioned in Chapter 19. For modeling purposes, it is common to work with "ice-equivalent" depth rather than measured depth. This is the depth along the ice core which would have been measured if the snow at the top of the ice sheet had been compressed into ice. Such theoretical compression shortens the D72 core by 17 meters.[2] Ice-equivalent depth was used in Figure B.1 because ice-equivalent depth was used in the data table from which the plotted data were obtained.

The mathematical transformation from measured to ice-equivalent depth does not compress the bottom of the ice sheet, which is already made up of ice. Only the upper layers are compressed. Consequently, 13 meters above the bedrock is still the point corresponding to the oldest post-Flood ice. The ice-equivalent depth of this point is (298.9 - 17 - 13 =) 269 meters.

The elapsed time between the top of the ice sheet and 269 meters ice-equivalent depth is found by calculating the integral:

$$\Delta t = \int_0^{269m} \frac{1}{\lambda}\, dh \tag{B.1}$$

where λ is the measured annual thickness at depth h.

This integral can be approximated in the straight line interpolation case shown in Figure B.1 by computing the sum:

$$\Delta t \approx \sum_{i=1}^{20} \frac{h_{i+1} - h_i}{(\lambda_{i+1} + \lambda_i)/2} \tag{B.2}$$

In this equation, the h_i's and λ_i's are the coordinates of the twenty points shown in Figure B.1. The points are enumerated with $i = 1$ at the top of the ice sheet, h_{21} is the ice-equivalent depth of the oldest post-Flood ice (i.e., 269 meters), and λ_{21} is obtained by a short linear extrapolation of points 19 and 20 to this depth.

This sum yields 5401 years. Subtracting the date of drilling (A.D. 1972) and allowing for the absence of year 0 in the B.C./A.D. calendar system yields 3430 B.C. as the best estimate of the the date for the 13 meter-above-bedrock ice.

---

[2] N. Reeh and W. S. B. Paterson, "Application of a Flow Model to the Ice-divide Region of Devon Island Ice Cap, Canada," *Journal of Glaciology* 34.116 (1988): 56.

Appendix C

Using Radiocarbon to Correct the Elk Lake Annual Layer Chronology

Anderson *et al.* report sixteen radiocarbon measurements on organic carbon from the Elk Lake cores (Table C.1).[1] We wish to use these measurements to estimate how many annual layers may have been left uncounted in the Elk Lake sediment cores back to the time of the Flood.

To use these samples correctly for the present purpose requires some knowledge of how radiocarbon dating works—the problem is actually not a trivial one. I will skip over the basics of the global production of radiocarbon and its subsequent distribution in various geophysical reservoirs for the sake of brevity and merely state that the most important point to be aware of in the present context is that radiocarbon dates will appear too old if carbon atoms which have not been derived from carbon dioxide in the atmosphere are taken up by the samples being dated. The taking up of such carbon is a common phenomenon in lakes because they commonly receive carbon atoms (as carbonates) from leaching of carbonate rocks and soil by rainwater runoff. Such carbon is referred to as old carbon, to distinguish it from carbon atoms derived from the atmosphere.

[1] Roger Y. Anderson, J. Platt Bradbury, Walter E. Dean and Minze Stuiver, "Chronology of Elk Lake sediments: Coring, sampling, and time-series construction," *Elk Lake, Minnesota: Evidence for Rapid Climate Change in the North-Central United States*, ed. J. Platt Bradbury and Walter E. Dean (Boulder: The Geological Society of America, Inc., 1993), 41.

Table C.1: Uncalibrated radiocarbon dates on samples from Elk Lake.

	Lab #	Layer #	^{14}C Age (yr B.P.)		
1	QL4018	0	420	±	60
2	QL4017	88	1160	±	65
3	QL1560	648	1420	±	80
4	QL1561	1100	2270	±	80
5	QL1562	2216	3360	±	70
6	QL1493	2317	3190	±	100
7	QL1563	2634	3370	±	70
8	QL1492	2666	3660	±	130
9	QL1564	2731	3510	±	90
10	QL1565	5084	5750	±	120
11	QL1494	5654	5290	±	100
12	QL1566	6694	7880	±	50
13	QL1495	7983	8550	±	140
14	QL1496	9061	9830	±	150
15	QL1497	10,500	11,380	±	180
16	QL1498	unlayered	17,000	±	800

The presence of old carbon in a sample gives it an old radiocarbon age even while it is still living. Fortunately, the problem can be corrected in many instances.

Elk Lake, which is located in calcareous glacial drift, is not exempt from the old carbon phenomenon. The trees which live around the lake do not contain old carbon because they derive all of their carbon atoms exclusively from the atmosphere. But anything which lives within the lake, such as fish or clams or diatoms, will contain old carbon. The radiocarbon dates on organic material from the Elk Lake cores will include an old carbon component, because the organic material comes from organisms which once lived in the lake.

The presence of old carbon in Elk Lake significantly complicates the task of determining what fraction of annual layers has gone uncounted on average, but, fortunately, it does not render it impossible. The old carbon phenomenon would be fatal to the present task only if the concentration of old carbon in the lake changed with time (for then it would be impossible to tell whether the change in measured radiocarbon concentration in the samples was due to the decay of radiocarbon atoms with time or to different concentrations of old carbon in the lake when the various samples were deposited).

It is very likely that the concentration of old carbon did change in the lake early in its history, because freshly deposited calcareous glacial till is likely to weather more rapidly at first. It is also probable that the concentration of old carbon in the lake would change when the vegetative cover surrounding the lake changed, because of the different rate of weathering of calcareous till likely to accompany such a shift. Thus, to keep the old carbon phenomenon from invalidating the radiocarbon estimate of missing layers, it is necessary to restrict the analysis to approximately the upper 3000 layers, where significant long-term changes in old carbon concentration in the lake would be unlikely.

Nine of the sixteen radiocarbon samples published by Anderson *et al.* (Table C.1) fall within this range. It is necessary to separate the constant old carbon contribution from the radiocarbon ages of these samples so they can be used to accurately measure real calendar years. There are two equations involved in this problem.[2] First is the relationship between the radiocarbon age of a sample which lived in the lake, $A_{LAKE}(t)$, and

[2]Gerald E. Aardsma, "Toward an Absolute Chronology at Elk Lake, Minnesota," *Radiocarbon* 38.3 (1996): 603–605.

that of a tree living at the same time beside the lake, $A_{ATM}(t)$. This relationship is expressed by the equation:

$$A_{ATM}(t) = A_{LAKE}(t) - A_{OLD} \quad \text{(C.1)}$$

In this equation, A_{OLD} is the old carbon contribution to the radiocarbon age of a sample which grew in the lake. It appears without functional time dependence, (t), in the equation because it is assumed constant over the time interval to which the equation will be applied, as just discussed.

The second equation provides an explicit relationship between the time variable, t, and the layer number, L.

$$t = -((1 + f)L + 23) \quad \text{(C.2)}$$

The 23 in this equation arises as follows. Since radiocarbon is the fundamental chronometric method in this case, it is most convenient to adopt the standard radiocarbon convention that $t = 0$ corresponds to A.D. 1950. (Thus, A.D. 1940 corresponds to $t = -10$, and so forth.) Now $L = 0$ corresponds to A.D. 1927, as mentioned above, which corresponds to $t = -23$.

The f in the equation represents the fraction of annual layers which has been missed in the counting process. It is the unknown to be solved for.

It is impossible to solve these two equations in closed form because the time dependence of radiocarbon age for samples which grew in the atmosphere (e.g., trees) does not correspond to any simple mathematical function. It is necessary to employ other means.

I have used the following approach. First choose a value for f and use it to calculate t for each of the nine radiocarbon samples. Then look up $A_{ATM}(t)$ in the 1993 *Radiocarbon* calibration issue.[3] Next plot $A_{ATM}(t)$ versus the measured $A_{LAKE}(t)$ given in Table C.1 in the "^{14}C age (yr B.P.)" column. According to equation C.1, this should yield a straight line with a slope of one for the correct choice of f. Finally, use a standard, unweighted linear regression to compute the slope and intercept for various choices of f.

Figure C.1 shows the resulting graph of slope versus f. A linear regression applied to these data yields a straight line which intersects $y = 1$ at $f = 0.176$. This says that on average 17.6% of annual layers

[3]Minze Stuiver and Bernd Becker, "High-precision decadal calibration of the radiocarbon time scale, AD 1950–6000 BC," *Radiocarbon*, 35.1 (1993): 57–65.

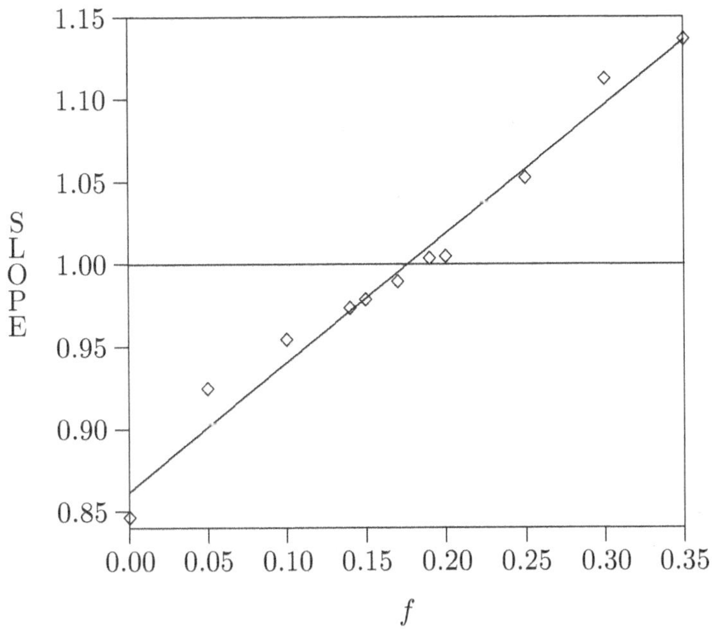

Figure C.1: Calculated slopes resulting from various choices of f. The diagonal line results from a linear regression applied to all of the data points shown.

were missed in the Elk Lake cores. When this value of f is substituted back into equation C.2 with $t = -(3520 + 1950) = -5470$ (i.e., the date of the Flood), the calculated layer number corresponding to the Flood is 4632.

Estimating the uncertainty in f determined in this manner is a little difficult. It is necessary to do so, however, to get some impression of the imprecision in the Flood layer number computed by this method. I have fit two other straight lines to selected data in Figure C.1 for this purpose. First, I have tried to estimate the smallest value f might reasonably be assigned by fitting just the data points $f = 0.05, 0.1$, and 0.3. This yielded $f = 0.154$ with the corresponding Flood layer at 4718. Second, to estimate the largest value of f, I have used just the two points $f = 0.0$ and 0.2. This gave $f = 0.195$ and the Flood layer at 4558. These results seem to be adequately summarized by the single statement that the Flood is expected to be seen, if present in the Elk Lake data, at layer number 4630 ± 90 (3σ).

The procedure followed in this appendix probably slightly overestimates the fraction of missed layers back to the Flood (and thus assigns the Flood layer range at layer numbers which are slightly too low). The analysis is necessarily restricted to the modern lake stage, but about 850 prairie lake layers are also involved in the calculation of the Flood layer. It seems probable that the fraction of layers which were missed by the researchers who did the layer counting was somewhat smaller during the prairie stage, because the annual layers are thicker on average during this stage and, hence, presumably harder to miss. It does not seem worthwhile to pursue this quantitatively, however. The correction which might result seems probably only a decade or two at best, and this seems adequately covered by the range of 180 layers which I have already allowed.

APPENDIX D

WARP1.FOR

The program listing below shows how I calculated the depth of the water of the Flood for various displacements of the inner core in the case of a PREM-like earth. That is, I have used PREM (the Preliminary Reference Earth Model), Appendix F of: Frank D. Stacey, *Physics of the Earth* (Australia: Brookfield Press, 1992) for most of the necessary geophysical parameters.

The code is in Fortran. I have executed it on a Windows DOS Pentium IV platform using gfortran via MinGW32 (specifically, tdm-gcc-4.7.1-2).

```
      program WARP1
!
!  WARP1 calculates the equipotential corresponding to the ocean surface
!    resulting from the inner core rising beneath the North Pole.
!  The calculation proceeds in degree increments from the South Pole to the
!    North Pole.
!  Results are saved in the output file WARP1.txt.
!  Provision is made at output for rotation of the equipotential about earth's
!    center.  The rotated result is saved in WARP2.txt.
!  Units are everywhere mks.
!  Coordinates are (y,z) with z along the South Pole -- North Pole axis.
!  The origin is at the center of the earth.
!
!  The inner core is treated as a point mass located at its own center.
!  The gravitational potential is calculated as the potential of a Preliminary
!    Reference Earth Model (PREM) earth having only a fluid core, plus the
!    potential of a ''mass anomaly'' due to the inner core mass point.
```

```
! The mass of the mass anomaly is calculated as a density anomaly times the
!   volume of the inner core.
! The density anomaly is the PREM density of the inner core at (0,0) minus the
!   PREM density of the outer core fluid at the displaced position of the inner
!   core, (0,z_ic).
! This approximation begins to break down for penetration of the inner core into
!    the mantle.  No correction is made for this, so limit z_ic to 2500 max.
!   z_ic must be positive, corresponding to displacement of the inner core
!   northward.
!
   implicit real*8 (a-h,k-z)     ! use double precision everywhere
   dimension c(0:180), s(0:180), h(0:180), area(0:180)
!
! User inputs
   z_ic = 2400.0D+3     ! z coordinate of the mass anomaly
   i_rotate = 25          ! angular distance (degrees) to rotate final equipotential
!
! Fixed constants and initial values
   M_e = 5.976D+24          ! mass of the earth
   pi = 3.14159265358979323846D+0
   delta_R = 1.0D+0          ! step size for equipotential search
   h = 0.0D+0
   i_pass = 1
!
! Calculated program constants
!
!    calculate density anomaly resulting from shifted inner core
!   PREM density at center of inner core
   rho_ic = 13088.48
!   polynomial for density of outer core fluid derived from PREM.
!       The PREM table was extrapolated into the inner core region by
!       subtracting from the inner core density at r the difference between
!       the inner core density and the outer core density at the inner core to
!       outer core boundary.  Then a polynomial was fit using Excel.
   c1 = 1.9713D-63
   c2 = 8.2814D-50
   c3 = 1.3526D-36
   c4 = 8.9719D-24
```

```
    c5 = 2.2880D-10
    c6 = 1.2493D+04
    zsq = z_ic*z_ic
    rho_oc = ((((-c1*zsq + c2)*zsq - c3)*zsq + c4)*zsq - c5)*zsq + c6
    delta_rho = rho_ic - rho_oc
!
!     calculate mass anomaly resulting from shifted inner core
    R_ic = 1.2215D+6          !  PREM radius for inner core
    Vol_ic = 4.0D+0/3.0D+0 * pi * R_ic**3
    M_a = delta_rho*Vol_ic
!
!     calculate depths of global ocean in Northern and Southern Hemispheres
!    There is greater water volume in the southern oceans than in the northern
!        oceans because there are more continents in the Northern Hemisphere
!    Correct for this by raising the ocean floor in the N and lowering it in the S
    R_e = 6.371D+6                !  PREM radius of earth to ocean surface
    Area_continents = 1.486D+14
    Area_Nc = 0.672D+0 * Area_continents
    Area_Sc = Area_continents - Area_Nc
    Area_e = 4.0D+0 * pi * R_e**2          !  surface area of PREM earth
    Area_oceans = Area_e - Area_continents
    Area_No = Area_e / 2.0D+0 - Area_Nc
    Area_So = Area_e / 2.0D+0 - Area_Sc
    Vol_oceans = 1.338D+18     !  National Geophysical Data Center estimate
    Vol_So = Vol_oceans * Area_So / Area_oceans
    Vol_No = Vol_oceans * Area_No / Area_oceans
!    These depths effectively give a modified PREM Crust--Ocean boundary
!        radius.  The water of the Flood sits on this base and fills to the
!        equipotential surface created by the displaced inner core.
    depth_N = Vol_No / (Area_e / 2.0D+0)
    depth_S = Vol_So / (Area_e / 2.0D+0)

!    In addition, continents will take up volume in the Northern Hemisphere
!        where the water is going.  Add this extra volume in to the volume
!        which must be accommodated inside the equipotential.
    Height_continents = 840.0D+0     !  ``Introduction to Geophysics'', Garland
    Vol_Nc = Area_Nc * Height_continents
```

```
!!  Calculated constant arrays
!     first calculate sin and cos at 1 degree increments
    do 10 i=0,180
        latitude = i - 90              !  earth latitude in degrees
        phi=latitude*pi/180.0D+0     !  phi = earth latitude in radians
        c(i)=cos(phi)
        s(i)=sin(phi)
10    continue
!    then calculate area of 1 degree latitudinal strips at surface of earth
    f1 = 2.0D+0 * pi * R_e
    breadth = f1 / 360.0D+0
    radius = breadth / 2.0D+0
    area(0) = pi * radius**2             !  S pole
    area(180) = area(0)                 !  N pole
    do 20 i = 1, 179
        length = f1 * c(i)
        area(i) = breadth * length
20    continue
    area(90) = area(90) / 2.0D+0     !  equator strip gets split N and S

!!  Volume loop
!    The volume of the trial equipotential is known to be too large initially.
!    Shrink the trial equipotential a little each loop until the volume is found
!        to be too small.

!  z0 is used to obtain a value for the trial equipotential.
!    It is the z coordinate of the trial equipotential at the South Pole.
!    It must be chosen to give an initial volume which is
!    known to be larger than needed to hold all the water of the Flood.
    z0 = R_e + 1.0D+0

30    write(*,*)'Height of trial equipotential above S Pole (m) = ', z0 - R_e

!  Calculate U0, the value of the trial equipotential (divided by -G).
    call pot(0.0D+0,-z0,z_ic,M_a,M_e,U0)

!  Find h of the trial equipotential for various latitudes, phi
!     start at height of equipotential at S Pole from previous iteration
```

```
    hguess = h(0)

    do 50 i = 0, 180
        deltay = delta_R * c(i) !  take 1 meter radial steps
        deltaz = delta_R * s(i)
        y = (R_e + hguess) * c(i)
        z = (R_e + hguess) * s(i)
        call pot(y,z,z_ic,M_a,M_e,U)
!     choose direction to step in search of the desired potential
        dir = 1.0D+0                    !  step outward
        if(U .lt.  U0) dir = -1.0D+0         !  step inward

!     step in or out until equipotential is crossed
40      y = y + dir * deltay
        z = z + dir * deltaz
        call pot(y,z,z_ic,M_a,M_e,U)
        if(dir*(U0-U) .lt.  0.0D+0) goto 40

!     calculate h(i), the height of the potential above sphere of radius R_e.
!        this will be inaccurate by at most 1 meter
        h(i) = sqrt(y*y+z*z) - R_e

        if(i .eq.  180) goto 50
!    now set hguess for next i iteration
        if(i_pass .eq.  1) then
!            On the first pass the shape of the equipotential is not known.
!            Use the h(i) just calculated as predictor.
            hguess = h(i)
        else
!            On subsequent passes use the old equipotential surface as
!                predictor.
            hguess = h(i+1)
        endif
50    continue
    i_pass = i_pass + 1

!  Calculate volume available for water inside trial equipotential
    volume = 0.0D+0
```

```
!    first do S
    do 60 i = 0, 89
!        calculate depth available for water above the ocean floor in the south.
        depth = h(i) + depth_S
!        depth does not extend into solid earth
        if(depth .lt.  0.0D+0) depth=0.0D+0
        volume = volume + area(i) * depth
60    continue
!    next do equator
!    calculate depth available for water above the ocean floor in the south.
    depth = h(90) + depth_S
    if(depth .lt.  0.0D+0) depth=0.0D+0
    volume = volume + area(90) * depth
!    calculate depth available for water above the ocean floor in the north.
    depth=h(90) + depth_N
    if(depth .lt.  0.0D+0) depth=0.0D+0
    volume = volume + area(90) * depth
!    finally do N
    do 70 i = 91, 180
!        calculate depth available for water above the ocean floor in the north.
        depth = h(i) + depth_N
        if(depth .lt.  0.0D+0) depth=0.0D+0
        volume = volume + area(i) * depth
70    continue

    Vol_excess = (volume/(Vol_oceans+Vol_Nc)-1.0D+0) * 100.0D+0
    write(*,*)'Excess volume percent inside trial equipotential = ', Vol_excess

!  Try shrinking trial equipotential surface by 1.0 meter at South Pole
    z0 = z0 - 1.0D+0

    if(volume .gt.  Vol_oceans+Vol_Nc) goto 30

!!  Save results
    open(unit=16, file='WARP1.txt')
    open(unit=14, file='WARP2.txt')
    do 80 i = 0, 360
        j = i
```

```
        if(i .gt.  180) j = 360 - i
        depthadj = depth_S
        if(j .gt.  90) depthadj = depth_N
        if(j .eq.  90) depthadj = (depth_N + depth_S) / 2.0D+0
!  calculate actual depth of water above PREM-like crust--mantle boundary
        depth = h(j) + depthadj
!        Depth does not extend into solid earth
        if(depth .lt.  0.0D+0) depth = 0.0D+0

!  save angle measured from S pole, equipotential depth, and water depth
!    Depths are in meters above mean sea level.
!    This is for graphing water depth versus degrees from S pole.
        write(16, fmt=''(i6, f7.0)'') i-180, depth-depthadj

!  save water depth above PREM crust in (y,z) coordinate system
!    This is for graphing water depth on scale diagram of earth.
        scale_factor = 350.0D+0     !  scale factor for graphing water depth
        R_water = (depth-depthadj+depth_S)*scale_factor + R_e
        j = i + i_rotate
        i_angle = j - 90
        if(i_angle .ge.  360) i_angle = i_angle - 360
        phi = i_angle * pi / 180.0D+0
        y_water = -R_water * cos(phi)
        z_water = R_water * sin(phi)
        write(14, fmt=''(f7.0, f7.0)'') y_water/1.0D+3, z_water/1.0D+3
80    continue
    close(14)
    close(16)
    write(*,*)'Done!'
    end

!*****************************************************
subroutine pot(y,z,z_ic,M_a,M_e,U)
    implicit real*8 (a-h,k-z)
!  Calculate total potential at (y,z) of entire mass distribution.
!  All ''potentials'' are divided by the negative of the gravitational constant
!    (i.e.  by -G) to speed up program execution.
!    calculate potential due to mass anomaly resulting from shifted inner core
```

```
      dz = z - z_ic
      U_a = M_a / sqrt(y*y + dz*dz)
!      calculate potential due to mass of earth less mass anomaly
      U_e = (M_e - M_a) / sqrt(y*y + z*z)
!      calculate the total potential at (y,z)
      U = U_e + U_a
      end
```

Appendix E

WARP1.TXT

Here is half of the WARP1.txt output file resulting from execution of WARP1.FOR for the user inputs shown in its listing in Appendix D. (I have deleted all of the initial WARP1.txt lines [for which the angles are negative] since the equipotential is symmetric about 0 degrees.) The left column gives angular distance from Iceland in degrees, and the right column gives the altitude of the surface of the Flood water in meters above today's mean sea level at that angular distance. Notice that the Flood has dropped below mean sea level by a distance of 80 degrees, and that the (idealized) ocean floor is exposed beyond 127 degrees.

This table is helpful in answering many questions arising in connection with the Flood. For example, suppose one wished to know whether anyone would have survived the Flood in Japan. The answer to this question obviously depends upon how deep the Flood was in Japan. The highest mountain in Japan is Mount Fuji. Its present elevation is 3776 meters. Now the elevation of the Flood waters at Mount Fugi is needed. Using a model globe of the earth and a piece of string it is easy to measure the angular distance from Iceland to Mount Fuji. I found roughly 77 degrees. From the WARP1.txt output table below the Flood altitude at 77 degrees is found to be only 277 meters. Clearly, Mount Fuji would have provided a place of escape from the Flood. In fact, most of Japan would have been safe from the Flood since much of the country is mountainous and well above 300 meters.

```
0 8884.
1 8880.
2 8870.
```

3	8853.
4	8829.
5	8798.
6	8761.
7	8717.
8	8667.
9	8610.
10	8548.
11	8479.
12	8405.
13	8325.
14	8240.
15	8150.
16	8054.
17	7954.
18	7850.
19	7741.
20	7629.
21	7513.
22	7393.
23	7270.
24	7144.
25	7015.
26	6884.
27	6751.
28	6615.
29	6477.
30	6338.
31	6198.
32	6056.
33	5913.
34	5770.
35	5625.
36	5481.
37	5335.
38	5190.
39	5045.
40	4899.

```
41 4754.
42 4610.
43 4465.
44 4322.
45 4179.
46 4036.
47 3895.
48 3754.
49 3615.
50 3477.
51 3339.
52 3203.
53 3068.
54 2935.
55 2803.
56 2672.
57 2542.
58 2415.
59 2288.
60 2163.
61 2040.
62 1918.
63 1797.
64 1678.
65 1561.
66 1445.
67 1331.
68 1219.
69 1108.
70 998.
71 891.
72 784.
73 680.
74 577.
75 475.
76 375.
77 277.
78 180.
```

```
79 85.
80 -9.
81 -102.
82 -192.
83 -282.
84 -370.
85 -456.
86 -541.
87 -625.
88 -707.
89 -788.
90 -867.
91 -945.
92 -1022.
93 -1098.
94 -1172.
95 -1245.
96 -1316.
97 -1386.
98 -1455.
99 -1523.
100 -1590.
101 -1655.
102 -1719.
103 -1782.
104 -1844.
105 -1904.
106 -1964.
107 -2022.
108 -2080.
109 -2136.
110 -2191.
111 -2245.
112 -2298.
113 -2350.
114 -2401.
115 -2451.
116 -2499.
```

117	-2547.
118	-2594.
119	-2640.
120	-2685.
121	-2729.
122	-2772.
123	-2814.
124	-2856.
125	-2896.
126	-2935.
127	-2974.
128	-2994.
129	-2994.
130	-2994.
131	-2994.
132	-2994.
133	-2994.
134	-2994.
135	-2994.
136	-2994.
137	-2994.
138	-2994.
139	-2994.
140	-2994.
141	-2994.
142	-2994.
143	-2994.
144	-2994.
145	-2994.
146	-2994.
147	-2994.
148	-2994.
149	-2994.
150	-2994.
151	-2994.
152	-2994.
153	-2994.
154	-2994.

```
155 -2994.
156 -2994.
157 -2994.
158 -2994.
159 -2994.
160 -2994.
161 -2994.
162 -2994.
163 -2994.
164 -2994.
165 -2994.
166 -2994.
167 -2994.
168 -2994.
169 -2994.
170 -2994.
171 -2994.
172 -2994.
173 -2994.
174 -2994.
175 -2994.
176 -2994.
177 -2994.
178 -2994.
179 -2994.
180 -2994.
```

Appendix F

Finding the Ark's Resting Place

The geographical region in which the ark must have come to rest has already been identified (Figure 28.1, page 221). This is a large region, but the probable resting place of the ark can be narrowed further by careful attention to the biblical narrative.

Narrowing the Search

Most importantly, notice that the biblical text demands the idea that the ark came to rest on a mountain. It cannot be supposed that the ark rested somewhere in an open plain because Genesis 8:4 states plainly that "the ark rested upon the mountains of Ararat." The broad setting is clearly mountains, not plain. But it also cannot be supposed that the ark came to rest in some valley in the mountains. The biblical narrative says that 72 days (Figure 16.3, page 119) elapsed after the ark had come to rest before the tops of neighboring mountains became visible. If the ark had landed in a valley between mountains, the mountains would have been visible long before the ark ever grounded. It is clear that the ark must have come to rest upon a mountain.

This narrows the search considerably, but not nearly enough. Ararat is a very mountainous region, and the bounded area of Figure 28.1 contains hundreds of mountains. Is there any way of narrowing the search any further? It seems possible to do so using a probabilistic approach. I proceeded as follows.

Initial Analysis

The basis of my analysis was Genesis 8:5,

> And the water decreased steadily until the tenth month; in the tenth month, on the first day of the month, the tops of the mountains became visible. (NASB 1975)

This is the first mention of visible land following the commencement of the Flood. I believe it contains several clues to the location from which Noah made this observation.

First, notice that this first mention of visible land is not of the mountain upon which the ark came to rest. I suppose it might be argued that Noah simply failed to mention the first sighting of the mountain the ark landed on, but this seems unlikely. Noah and his family had watched the waters of the Flood swallow the land—their homeland. Subsequently they had drifted upon a boundless and seemingly relentless ocean for many months. Surely the first appearance of land as the Flood began to wane would have been a sight of great significance to those aboard the ark. It seems inevitable that the first sighting of land would have been recorded by Noah.

So the fact that the first land Noah reports seeing is the neighboring mountains implies that he had not seen the mountain the ark had come to rest upon by the time the ark grounded, or even by the time the neighboring mountains emerged from the water 72 days later.

This deduction suggests that the mountain the ark grounded on must still have been hidden beneath the water when the ark grounded upon it. If this were not the case, then Noah should have seen the mountain as the ark drifted toward it. And in that case, that sighting should have been the first land reported by Noah, rather than the sighting of the neighboring mountains.

But if the mountain was still hidden beneath the water when the ark grounded upon it, then it must be the case that the ark came to rest at or near the summit rather than on the side of this mountain.

Thus Genesis 8:5 leads to a picture of the ark having come to rest at the summit of its mountain. When it first grounded, all that could be seen out to the horizon in all directions was water. As the water receded day after day the horizon would have moved farther and farther out away from the ark. Finally, after 72 days, Noah saw the tops of neighboring mountains poking through the surface of the water for the first time.

The most important clue to Noah's location to be found in this verse is that no neighboring mountains were visible when the ark grounded at the summit of its mountain. This would not be possible in the case of many mountains. If, for example, the ark had come to rest on a mountain having a near neighbor twice its height, then the near neighbor would have been sighted *before* the ark had come to rest, not 72 days after. Clearly, this clue can be used to eliminate many mountains in the Ararat region from further consideration as candidates for the mountain the ark landed on.

It might be thought that this idea, taken to its logical conclusion, leads immediately to the result that the ark must have landed on the highest mountain in the Ararat region. If it had landed on a shorter mountain, wouldn't the taller mountain already have been exposed above the surface of the Flood waters? In fact, because of the curvature of the surface of the spherical earth, it would have been possible for some mountains, which had already emerged from the Flood waters, to be hidden from Noah's view behind the horizon. (See Figure F.1.) So Genesis 5:8 does not lead inevitably to the conclusion that the ark came to rest on the tallest mountain in the Ararat region. But the fact that it took 72 days for neighboring mountains to be exposed, during which time "the water decreased steadily," does suggest that the mountain the ark had landed on was the tallest one in its general vicinity.

The Clue Applied

I applied this clue to the mountains of the Ararat region as follows.

I purchased five Tactical Pilotage Charts (TPCs) which together covered the area of interest to the present study.[1] These are topographical maps having a scale of 1:500,000. They are intended for aircraft pilots and are considerably more detailed and accurate than maps found in ordinary world atlases.

[1]The five maps I worked from were F-4C, F-4D, G-4A, G-4B, and G-5A. These were all prepared and published by the Defense Mapping Agency Aerospace Center, St. Louis, Missouri. They can be ordered from: NOAA Distribution Branch (N/CG33), National Ocean Service, Riverdale, MD 20737-1199. I obtained my copies through Map World, 123-D North El Camino Real, Encinitas, CA 92024, (800)246-6277.

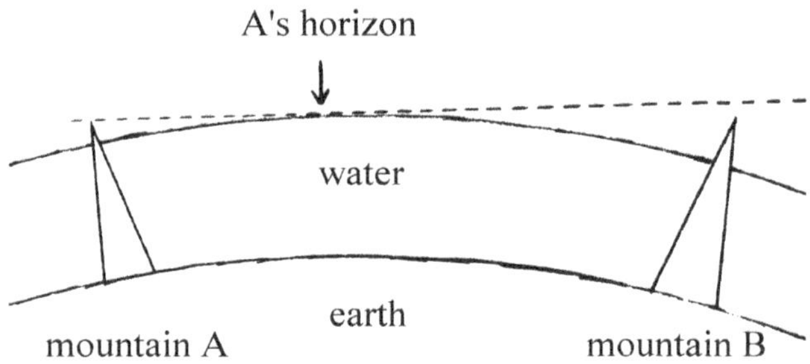

Figure F.1: Though mountain B is taller than mountain A, it is hidden behind A's horizon. Nothing below the dashed line is visible from A past the horizon.

Uncertainties

I must pause and give a word of caution before proceeding. Even though these maps are very good, they are not perfect. In the present study, it would have been preferable for the heights of all mountain peaks to be known to better than ± 50 feet (3σ). But these maps do not provide such high precision. It is a general rule of thumb in science that a 1% uncertainty is fairly good for most physical measurements. Many of the mountains in the Ararat region are over 10,000 feet high. It was no surprise, therefore, to find noted on the TPC maps that the standard uncertainty in elevations marked with a • on the maps "does not exceed 100 feet" while those marked with an × can have an uncertainty two or even three times as large.

But even if these maps could give elevations accurate to the nearest foot, the present study would still face some uncertainty. The fact is that mountains do change with time, and 5500 years (back to the Flood) is a long time. Many of the mountains in the Ararat region—including Mount Ararat—are volcanic. There is no guarantee that none of these mountains has gained or lost substantial height through volcanic activity since the Flood.

Obviously, it is necessary either to be (intelligently) prepared to put up with these uncertainties, or to forgo the present investigation altogether. But there is no reason to be timid, in my opinion. While it is clear from the outset that the final results of the present search cannot be guaranteed certain, the fact is that the mountains of Ararat *have* been measured to reasonable precision, and it is reasonable to believe the topography of these mountains today mirrors the landscape Noah looked out upon from the ark 5500 years ago to a very large extent. Furthermore, it is possible to minimize the impact of these uncertainties by adopting a probabilistic approach to the problem which considers multiple factors together. I have adopted such an approach in the current study.

But most importantly, please note that I am not attempting to prove anything here. I am, first of all, not attempting to prove that Genesis is true when it says the ark landed upon the mountains of Ararat. I am also not attempting to prove that the ark landed upon any given mountain in Ararat. Rather, I am only attempting to determine as scientifically as possible which of the many hundreds of mountains in the Ararat region seems at the present time to satisfy best the requirements of the biblical narrative. For this purpose, the TPC maps seem adequate, even if not entirely ideal.

On With the Search

Many spot elevations are given on the TPC maps. These are predominantly notations of the highest points—of obvious concern to pilots. Most simply note the heights of the peaks of mountains. A few, in low-lying areas, appear to give the elevations of small hills. Because the mountain the ark rested on is expected to be the tallest in its vicinity, the height of that mountain is expected to be included among the spot elevations on these maps.

The first step in my search procedure was to test every mountain for which a spot elevation was given on these maps, in the region of interest to the present study, to see if any other mountain would be visible from it when it was barely covered by water. Because of the large number of mountains involved, I found it necessary to use a computer to carry out this test. This meant that the positions (longitude and latitude) and heights (i.e., spot elevations) of the mountains shown on these maps needed to be entered into the computer. To avoid any subjective bias, I decided to enter every spot elevation given on the maps within the chosen area of study, even those few which obviously corresponded to just low hills. I entered a total of 1441 spot elevations with their corresponding position coordinates from the region of study.

As mentioned above, I chose as the region of study the area bounded by longitudes 37°E on the west and 48°E on the east, and latitudes 36°N on the south and 42°N on the north. (See Figure 19.1, page 136.) The extent of the studied region was deliberately chosen to exceed the extent of the Ararat region because candidate mountains within the Ararat region must still be tested against neighbors in areas adjacent to the Ararat region.

I next had to program the computer with the search criterion. This criterion was made suitably quantitative as follows.

The question of whether a neighbor mountain B will be visible from candidate mountain A when A is barely covered by water divides into several cases. If the elevation of B is less than A, then B will still be covered by water and will not be seen from A. If the elevation of B is greater than A then B will be out of water. It will be visible from A if it is close enough to A. It will be hidden behind the horizon if it is too far from A.

A little math quickly reveals that the critical distance, r, at which B will be barely visible from A is given by the equation:

$$r = R \arccos \frac{R}{R+e} \tag{F.1}$$

In this equation, e is the difference in height of the two mountains, A and B, and R is the mean radius of the earth, for which I have used the value 6371 km. This equation shows, for example, that if B is 1 km higher than A, then it will be visible from A when A is barely covered by water if the distance between A and B is less than 113 km.

I wrote a simple computer program which uses this equation to test each mountain in the region of study against all other mountains in the region of study to see which mountains would have no visible neighbors when barely covered by water. Of the 1441 peaks entered, only 41 survived this test.

More Clues Applied

But 41 candidates is still 40 more than desired. To reduce the number of candidates further, I evaluated each of these 41 candidates in six categories.

My strategy at this point is perhaps best explained by an analogy. Imagine the police are looking for a certain criminal. They have rounded up 41 bad guys and are reasonably confident the man they are looking for is one of these 41. Unfortunately, the description of the criminal they have to go on is rather abbreviated, and worse still, they have been tipped off that the criminal may have had his appearance altered since the crime. Nonetheless, they determine to do the best they can using six characteristics of the criminal which they have deduced from the scene of the crime: his height was about 5 feet 7 inches; his hair was brown; he seems to have been muscular; ...

The police detective ranks each bad guy in each of these six categories. For example, in the height category, the bad guy whose height is closest to 5 feet 7 inches is given a rank of 1. The bad guy whose height is next closest gets a rank of 2, and so on.

Clearly, if bad guy X is found to rank first in all six categories while the ranks of the other bad guys show no consistent behavior in the six categories, then the detective should choose X—he is an excellent match to the description.

But the detective doesn't need to have such clear-cut evidence to make a good choice. He can use the ranking he has done to evaluate the probability a given bad guy is the man he is looking for.

To carry out this probability test, the detective converts the rank numbers to probabilities by simply dividing the ranks by the number of bad guys. For example, bad guy Y ranks 37 in the height category with a height of 5 feet 1 inch. From this the detective determines that the probability of one of these 41 bad guys having a height within 6 inches (i.e., 5 feet 7 inches minus 5 feet 1 inch) of the expected height of 5 feet 7 inches is 37/41.

The detective evaluates the relative probability of a given bad guy being the man he is looking for by multiplying the six probabilities for that bad guy together, one from each category. By doing so, he is evaluating the total probability of that bad guy looking as similar as he does to the description of the criminal. A relatively high probability of looking that similar means lots of bad guys look that similar and this is probably not the bad guy he is looking for. A relatively low probability, however, means it is unlikely for a bad guy to look that similar to the description of the criminal, and in that case the degree of similarity may be more than just a coincidence.

If the detective finds that bad guy X has a total probability (after multiplication of the six probabilities) which is very much lower than the probability for the other 40 bad guys, then he has objective grounds for once again choosing X. X is then, by far, the most likely of the lot.

Note, however, that the detective will not be able to make a choice if two or more bad guys share nearly identical lowest probabilities. This would mean they were equally similar to the description. (Perhaps the criminal has a twin brother, also pursuing a career in crime!) If none of the bad guys has a probability much lower than any of the others, then the test fails altogether, and the detective is back to the drawing board. This would perhaps suggest that the criminal did a truly remarkable job of altering his appearance before the test was conducted.

Now I will divulge the information that Mount Ararat is among the 41 candidates which are about to be examined by the procedure just described. But to protect the innocent (actually, to keep the discussion unprejudiced), other candidates will remain anonymous until the completion of the examination. At this stage, it is obviously important to minimize subjective bias and maximize the degree of conformity of the test to the Genesis narrative.

Category 1

I first looked at the distance from each candidate to the heart of the Ararat region around Lake Van. Some of the candidates were out near the border of the region of study, in areas which almost certainly were never part of the Ararat region—to the west of Malatya, in the plains of Syria to the south, and to the east of Lake Urmia, for example. Because of their location outside the Ararat region, these obviously have very little likelihood of being the true mountain the ark rested on.

To proceed quantitatively, I defined the "heart" of the Ararat region at longitude 43°E and latitude 38.667°N. This, according to the TPC maps, falls in the middle of Lake Van, around which, according to the scholars, the kingdom of Ararat was centered. I then wrote another computer program to determine the distance between this point and each candidate mountain. The candidates were then ranked. The candidate which was closest to the heart of Ararat was assigned a rank of 1, and the one furthest from the heart of Ararat was assigned a rank of 41.

Category 2

I next computed a rank for each of the 41 candidates based on the height of the candidate relative to its first seen neighbor. The idea here is that Genesis records that 72 days elapsed between the grounding of the ark and the appearance of neighboring mountains. During these 72 days the "water decreased steadily." Meanwhile, in just 160 days from the time the ark came to rest the ground around the base of the mountain was observed by Noah to be dry. This would mean that the mountain the ark was on stood (160/(160-72)=) 1.8 times higher above its base than the height of the water when the first neighbor appeared, if the water decreased at a constant rate. While it is unlikely the water decreased at a constant rate for this entire interval, the indication still seems fairly strong that the mountain the ark came to rest upon was significantly taller than its neighbors.

I calculated the height of each of the 41 candidate mountains above its base and computed the ratio of this height to that of the height of the water above this same base when the first neighbor mountain was seen. I defined the base level as the height of the water at which roughly 90% of dry land today within a 50 mile radius of the candidate mountain would be above water back then. I obtained my estimate of this base level by visual inspection of the contours on the TPC maps around each candidate.

The candidate mountain having the highest ratio achieved the rank of 1; the candidate with the lowest ratio (equal, in fact, to 1) was ranked 41.

Category 3

For the third category I compared the water heights when the first two neighboring mountains became visible from each candidate. Genesis 8:5 reports neighboring "mountains" (plural) as the first visible land. This means that at least two mountains became visible during an interval of 24 hours.

To measure the relative conformity of each candidate to this biblical requirement, I computed the difference between the height of the water when the first neighbor was seen and the height of the water when the second neighbor was seen. I then divided this by the difference between the height of the candidate mountain and the height of the water when the first neighbor was seen.

For a constant rate of decrease in water height, this ratio should be less than 1/72—1 being the maximum number of days allowed to expose the second neighbor after the first has become exposed, and 72 the number of days it took to expose the first neighbor from the time the ark grounded. The closer this ratio is to zero, the more likely it is that the two neighbors would have been seen at the same time.

The candidate having the lowest ratio was assigned rank 1 in this category.

Category 4

In the fourth category, I looked at the shape of the 41 candidate mountains. As noted above, Noah seems to have been unable to see the mountain the ark rested upon from inside the ark. This is presumably because his view much below the horizontal plain was restricted by the upper deck of the ark itself.

Such a restricted view is implied in several ways by the Flood narrative: the fact that Noah had to remove the covering of the ark to see that the ground was dry around the base of the mountain, and the fact that Noah had to send birds out from the ark to learn about his immediate environs, for example.

But this restricted view must be coupled with the idea that the mountain the ark had landed on must have been somewhat conical and steeply sloped to explain Noah's actions and observations fully. Clearly, the narrower the top of the mountain the ark rested upon, the more probable it becomes that Noah would have been unable to see it from inside the ark.

I calculated the ratio of the height of the candidate mountain exposed above water when the first neighbor became visible to the maximum width of the exposed portion of the mountain. The width was measured horizontally from the summit to the widest point on the exposed mountain. This widest point was estimated from the TPC maps. I used the closest contour to the height of the water wherever possible. With candidate mountains having a first-visible neighbor of nearly the same height, it was necessary to use the next lower contour to estimate the shape. In a few cases, it was necessary to go down one or two contour increments (usually 500 or 1000 feet) on the mountain to reduce measurement error.

The candidate with the highest ratio of height to width received rank

1 in this category.

Category 5

Genesis 8:4 says "the ark rested in the mountains of Ararat." For Category 5, I ranked the 41 candidates according to how "in the mountains" they could be said to be.

The TPC maps show that the Ararat region is made up of some large lakes, a significant proportion of high plateau, a few sizable valleys, and extensive foothills in addition to some large mountainous areas. Mountains are conspicuous on the TPC maps because elevations above 9000 feet are colored differently than lower elevations, and many of the mountains of Ararat exceed this elevation.

To estimate how "in the mountains" a candidate was, I had the computer count the number of spot elevations greater than 9000 feet within a 50 mile radius of the candidate. The candidate with the greatest number of such tall neighbors was given the rank of 1.

Category 6

I took my final clue from Genesis 11:2. It says of Noah's descendants, "And it came about as they journeyed east, that they found a plain in the land of Shinar and settled there" (NASB 1975). This seems to imply that the discovery of Shinar by Noah's descendants was accidental rather than deliberate.

It seems probable that Noah's descendants would simply have followed the course of a river as they journeyed. This would ensure them adequate water while in the Ararat region, which receives only light seasonal rain today. As they moved farther east and south, it would have become increasingly mandatory to follow the river since it seldom rains at all in those regions.

The numerous mountain streams of the southern Ararat region generally find their way ultimately into the Tigris. It seems probable that Noah's descendants happened upon Shinar simply as a consequence of following some such stream to the Tigris and then following the Tigris to the east.

In any event, it seems reasonable to suppose that the closer the ark landed to the Tigris, and the farther it landed down the Tigris toward

Shinar, the more likely Noah's descendants would be of winding up in Shinar.

The Tigris leaves the designated area of study at longitude 43.33°E and 36.0°N. I had the computer rank the candidates according to their proximity to this point. The closest candidate was given the rank of 1.

Results and Discussion

Table F.1 shows the ranking of each candidate in each category. The column on the far right shows the relative likelihood that the corresponding candidate is the mountain the ark landed upon, according to the six criteria outlined above.[2] These likelihoods are relative to Candidate 1, which is Mount Ararat.

It is apparent from Table F.1 that the traditional claim that Mount Ararat is the resting place of the ark is not supported. There is another candidate—Candidate 18—which is considerably more likely to be the mountain the ark rested upon than Mount Ararat. According to this study, Candidate 18 is 62.4 times more likely than Mount Ararat to be the mountain which corresponds to the biblical narrative.

Candidate 18 is labeled "Cilo Dagi" on the maps; I will call it Mount Cilo. Its position is shown in Figure 28.1 (page 221).

Table F.1 shows that Mount Ararat (Candidate 1) and Mount Cilo (Candidate 18) both have high ranks in the first four categories. They are about the same distance from the center of Lake Van. They are both quite tall relative to their neighbors. They both have two first-visible neighbors of comparable height. And they are both steeply sloped. Notice, however, that Mount Cilo ranks higher than Mount Ararat in three of these four categories.

Though Mount Cilo is already more likely than Mount Ararat to be the mountain the ark landed on when just these first four categories are considered, it is the final two categories which separate them most distinctly. Mount Ararat ranks thirteenth for being "in the mountains" while Mount Cilo ranks third. A glance at the TPC maps confirms that this relative ranking is correct. Mount Ararat isn't really "in the mountains." A broad valley, associated with the Aras River, runs all along its northeast side, for example. This valley obtains a width of some 50 miles

[2]The relative likelihood was computed by dividing the product of the six rankings for Mount Ararat (Candidate 1) by the product of the six rankings for each candidate.

Table F.1: Ranks and relative likelihoods for 41 candidate mountains.

candi-date	category						relative likelihood
	1	2	3	4	5	6	
1	6	2	8	2	13	18	1.000000
2	1	14	35	24	10	10	0.038204
3	4	18	6	38	13	17	0.012384
4	8	19	23	25	5	22	0.004673
5	31	26	36	41	41	12	0.000077
6	17	20	7	34	41	14	0.000967
7	26	32	27	14	41	21	0.000166
8	22	41	22	27	41	20	0.000102
9	24	28	32	3	10	29	0.002401
10	18	6	10	35	10	28	0.004245
11	11	41	2	26	28	16	0.004276
12	7	13	11	11	25	15	0.010881
13	25	5	3	37	41	7	0.011282
14	3	34	39	12	24	8	0.004902
15	14	12	16	31	41	1	0.013151
16	2	32	30	9	1	5	0.520000
17	10	41	2	29	24	9	0.008747
18	5	4	4	1	3	3	62.400000
19	13	21	25	7	15	2	0.031347
20	40	10	13	39	41	25	0.000216
21	41	34	33	36	41	31	0.000021
22	29	41	40	18	41	26	0.000049
23	36	3	17	32	28	32	0.000824
24	39	41	31	40	41	38	0.000015
25	33	28	19	16	17	37	0.000254
26	30	41	41	33	41	6	0.000110
27	28	32	18	22	30	4	0.001055
28	21	26	38	4	20	11	0.002461
29	12	17	15	17	4	19	0.011364
30	16	24	22	10	2	24	0.011080
31	32	1	9	20	24	23	0.014130
32	34	9	12	19	18	13	0.002752
33	23	16	5	28	15	30	0.001938
34	38	8	28	8	26	40	0.000634
35	37	15	14	5	19	41	0.001485
36	20	32	27	15	17	34	0.000300
37	9	22	24	21	10	27	0.001667
38	19	35	34	6	13	35	0.000728
39	15	7	20	13	6	33	0.008312
40	27	12	29	23	24	36	0.000241
41	35	23	37	30	27	39	0.000048

immediately to the north of Mount Ararat. An extension of this valley wraps around the base of Mount Ararat and its sister, Lesser Ararat, all along the south.

Mount Ararat and its sister are really somewhat isolated from other mountains. Mount Ararat is a very impressive mountain in photographs for just this reason. It is an enormous mountain, and it sits pretty much out in the open, dominating the view from the extensive valleys and plains at its base. But also, for just this reason, it is not really appropriate to say "the ark rested in the mountains of Ararat" if the ark grounded at the summit of Mount Ararat. It seems that a more accurate description in that case would be "the ark rested upon **a mountain** in Ararat." The setting of Mount Ararat doesn't seem to match the biblical description very well—which is why its rank is only 13 in this category.

By way of contrast, Mount Cilo really *is* "in the mountains." It is in the middle of a very mountainous area, and many of the mountains which surround it are over 10,000 feet high. Its high rank of 3 clearly reflects the good match between Mount Cilo's setting and the biblical description.

Mount Ararat is also located much farther from the Tigris than Mount Cilo. Any stream followed from the base of Mount Ararat leads to the Aras River. Following the Aras River leads eventually to the Caspian Sea. There is no obvious reason why anyone would come upon the land of Shinar starting from Mount Ararat.

In contrast once again, every stream which flows from the base of Mount Cilo leads ultimately to the Tigris and to the land of Shinar.

Appendix G

Depth of the Flood at Mount Cilo Versus Time

The Data

Day 0

The first point of interest for constructing a graph of the depth of the Flood versus time is Day 0, immediately preceding the start of the Flood. The depth which we need to know to be able to graph this point is simply the mean sea level back at that time.

Richard G. Fairbanks displays a graph of mean sea level which includes the period in question.[1] It is based on radiocarbon dated corals. It shows that the sea level was 5 ± 1 meters (1σ) below the present mean sea level 5,500 years ago.

Day 40

The text repeatedly mentions the initial forty days and nights of the Flood.[2] This period was one of rapidly increasing water depth until the rise of the inner core was stopped by the mantle on Day 40. At that point, the geoid warp caused by displacement of the inner core ceased to

[1]See Figure 2 (p. 639) of: Richard G. Fairbanks, "A 17,000-year glacio-eustatic sea level record: influence of glacial melting rates on the Younger Dryas event and deep-ocean circulation," *Nature* 342 (7 December 1989): 637–642.

[2]Genesis 7:4,12,17.

grow, and flooding would soon have stabilized at full depth. The actual depth of the water on Day 40 (and its associated uncertainty) is obtained from the Day 150 data below.

Day 150

The single most major piece of quantitative data regarding the depth obtained by the Flood is the one which has been used extensively already, found in Genesis 8:4. There we are told that "the ark rested upon the mountains of Ararat." This records the grounding of the ark near the summit of Mount Cilo on Day 150 of the Flood, or very shortly thereafter, as previously discussed on page 121. The implication is that the water of the Flood was just deep enough barely to cover Mount Cilo on Day 150.

Mount Cilo's summit presently stands at 13,566±33 feet (1σ) above mean sea level.[3] Questions of the possible subsidence or elevation of the Ararat plateau and of erosion of the mountain since the Flood suggest that the depth of the Flood on Day 150 should probably be regarded as uncertain to within 500 or 600 feet (3σ) of the present height of Mount Cilo. I therefore estimated a 1σ uncertainty of ±200 feet (±60 meters). Rounding the measured elevation up as previously (page 260) yields 4200±60 meters above mean sea level as the depth of the Flood on Day 150.

Day 222

Genesis 8:5 records that the water decreased steadily after the ark had grounded, and that on Day 222 the tops of neighboring mountains became visible.

Mount Cilo is surrounded by tall mountains. The tallest near neighbor is located about 25 miles (40 km) to the northeast. Its elevation is given on TPC map G-4B as 12,493±100 feet (3σ).[4] By the time the water of the Flood had decreased to this depth, the summit of Mount Cilo would have been (13,566 - 12,493 =) 1,073 feet (327 m) out of water. According to equation F.1 (page 350), the horizon would have been out at

$$r = 6371 \times arccos[6371/(6371 + 0.327)]$$

[3]Map TPC G-4B, Defense Mapping Agency Aerospace Center, St. Louis, Missouri.

[4]Map TPC G-4B, Defense Mapping Agency Aerospace Center, St. Louis, Missouri.

which equals 65 kilometers (about 40 miles) by that point. Thus this tallest neighboring peak would have been inside the horizon as viewed from Mount Cilo on Day 222 when it first breached the surface of the water. Therefore, the depth of the water on that day would have been just equal to the height of the peak of this neighboring mountain. This yields the data point 3800±60 meters for Day 222.

Day 310

The determination of the depth of the Flood on Day 310 is more dependent on interpretation of the Genesis narrative (and hence, more uncertain) than the previous data points have been. We learn from Genesis 8:13 that "the water was dried up from the earth" on this day. The interpretive question arises over whether this is intended to be understood in an absolute or a relative sense. Did Noah intend to convey that the region around the mountain over which he looked after he had removed the covering of the ark on Day 310 had come to a terminal state of dryness (similar to what one would see there today), or did Noah mean that the scene was dominated by dry land rather than by water?

Genesis 8:7 may offer some help with this interpretive difficulty. It informs us that the raven which Noah had sent out of the ark "flew here and there until the water was dried up from the earth." Whitcomb and Morris make an interesting connection between Genesis 8:7 and Genesis 8:13, which describes the state of dryness on Day 310.[5] The phrase "the water was dried up from the earth" appears in both verses. Apparently, then, the raven was seen flying about until about Day 310.

I am not an expert on the behavior of ravens, but it is clear that the raven is not a sea-going bird. It seems reasonable to suppose that the water surrounding Mount Cilo would constitute somewhat of a barrier to the raven, holding the raven in the vicinity of the ark as long as it was high enough. As the depth of the water decreased and neighboring mountains began to appear, it seems likely the raven could have flown away from Mount Cilo to these other mountains if it had wished to do so. But there is no obvious reason why it would have wished to do so as long as its food supply around Mount Cilo was adequate—which seems likely following such a Flood. Once the water had decreased to the point

[5]John C. Whitcomb, Jr. and Henry M. Morris, *The Genesis Flood* (Philadelphia: The Presbyterian and Reformed Publishing Company, 1961), 8, Figure 2.

that a flight of several miles over open water was no longer required for the raven to leave the vicinity of the ark, however, it seems probable that the raven would have strayed away from that vicinity in its daily search for food.

If this speculation on the probable behavior of the raven has merit, then it suggests that the raven would certainly have wandered away by the time the water had decreased to 5,750 feet. At that depth, Mount Cilo would no longer have been an island. By this point, for example, a person would have been able to walk, completely on dry ground, the 25 miles to the first neighbor which had appeared several months previously. But the raven was probably still confronted by somewhat of a water barrier when the depth of the water was 7,000 feet. At that depth, the raven would still have had to cross a two- or three-mile stretch of water at the narrowest point to leave the vicinity of Mount Cilo.

These considerations lead me to propose a probable depth of 2000±60 meters (6,600±200 feet) on Day 310.

Day 365

Genesis 8:14 records simply that on Day 365 "the earth was dry." At that point Noah and the other occupants of the ark disembarked. One gets the impression that things were pretty much back to normal around Mount Cilo by that point. This suggests a water depth no greater than about 3,500 feet, the approximate altitude of the river valleys in the vicinity of Mount Cilo today.

The text does not rule out the possibility that the water had receded much further than this by that point, however. It seems possible that the water may even have returned to its normal, pre-Flood state by Day 365.

An estimate which takes both of these extremes into account is 500±200 meters (1σ).

The Graph

The six data points discussed above are shown in the table at right. Figure 38.1 (page 279) displays a graph of these points.

Day	Depth (meters)	Uncertainty (1σ)
0	-5	± 1
40	4200	± 60
150	4200	± 60
222	3800	± 60
310	2000	± 60
365	500	± 200

Table G.1: Depth of the Flood data in meters above mean sea level today.

INDEX

AARDSMA
RESEARCH & PUBLISHING

Also by Dr. Aardsma

A New Approach to the Chronology of Biblical History from Abraham to Samuel

The Exodus Happened 2450 B.C.

Available through Aardsma Research & Publishing
412 N Mulberry St.
Loda, IL 60948-9651.

Order on-line through www.BiblicalChronologist.org.